NELSON'S

HORSEMASTER SERIES

# British Native Ponies and their Crosses

PHYLLIS HINTON

# British Native Ponies and their Crosses

NELSON

THOMAS NELSON AND SONS LTD
36 Park Street London W1Y 4DE
PO Box 18123 Nairobi Kenya

THOMAS NELSON (AUSTRALIA) LTD
597 Little Collins Street Melbourne 3000

THOMAS NELSON AND SONS (CANADA) LTD
81 Curlew Drive Don Mills Ontario

THOMAS NELSON (NIGERIA) LTD
PO Box 336 Apapa Lagos

THOMAS NELSON AND SONS (SOUTH AFRICA) (PROPRIETARY) LTD
51 Commissioner Street Johannesburg

This book first published 1971

Printed in Great Britain by
Cox & Wyman Ltd
London Fakenham and Reading
0 17 147203 9

# CONTENTS

# ILLUSTRATIONS

*(between pages 56 and 57)*

*All photographs courtesy of John Nestle*

## Chapter I

# PONY POTENTIAL

To appreciate fully the remarkable influence of our British native ponies on riding today it would be necessary to travel the world and spend quite an appreciable length of time in a large number of different countries. Equally, to give a worthwhile history of these remarkable ponies, years of careful and wisely directed study, the examination of records, and a considerable number of pages in which to describe how each breed came into being, its background and present potentialities, would be but a part of such an encyclopedic undertaking.

In this less ambitious and possible more practical book I have tried to describe adequately each breed and to indicate its most important characteristics, or assets, if you wish. I have referred to the care which may help to keep the ponies fit and happy; given, I hope, some indication of the various types of pony which have been bred from the pure stock and I have tried to explain how this was done (i.e. the way in which they have been crossed). Finally, we examine the practical use and value of these crosses.

From the practical angle the pony is the thing of the moment as they are comparatively easy to care for and can be used in a variety of ways, from showing to competitive events and of course, for breeding. Additionally, there is a good export market for the right pony.

From them can be bred such a variety of types it might be well to give some idea of their scope. First of all, the pure-bred registered pony will accomplish successfully many different jobs, depending up to a point on his size and ancestry, i.e. what he has been bred to do. He will also be in demand to start studs overseas and there he may be crossed with quite a startling variety of breeds indigenous to the country of his adoption which may include both ponies and horses.

Before starting a stud, whether it be a big one or just a small experiment, it is of very real importance to decide what *type* of pony you want to breed, what it must be able to do and what market exists for it. Haphazard, casual breeding without pattern or sequence leads to the production of mediocre ponies without a future and will not improve your bank balance or your reputation. Ponies are wanted for so many different things and it is important to realize the qualities expected by your prospective purchasers and the type in which they are interested.

For instance, the show pony in Britain is in many ways a very separate and exclusive animal – far more so than in Canada, for instance, where he is expected to be versatile and generally useful as well as beautiful – or perhaps that is not an entirely accurate definition. It might be better to say that a different type of performance is expected over here, although one wonders whether the Canadian idea may not eventually become increasingly popular in this country.

Let us examine in detail the model or show pony which competes in the not over 12·2 h.h., or 13·2 h.h., and 14·2-h.h. classes each season and what the judges look for in this pony, remembering that its counterpart in the States and in Canada will probably have to compete in a considerable number of performance classes in addition to model classes all on the same day. This may help us to get a clearer picture of what to aim at if we decide to breed for either of these markets.

For a number of very sound reasons the British Show Pony Society has issued various rules, and exerts an appreciable control over the showing of ridden ponies in this country, just as the National Pony Society and the Ponies of Britain deal with in-hand classes. The B.S.P.S. lays down the standard of manners and the paces required in the ring, and does not fail to emphasize its ruling in the case of the smaller ponies (not over 12·2-h.h., ridden by children not more than 12 years of age), together with many other rules, including details of measurement. Notes on its annual handbook are given in Chapter 7

However, most of us know that all the ponies, irrespective of size, should have a fair standard of good manners, suit the young rider (i.e. the age group which they are expected to carry) and be prepared first to walk, trot and canter round the ring, keeping behind each other, or passing to take the lead if the rider so decides. They must alter pace at the demand of the ring steward, line up on request, go to the front and give an individual show, stand to be examined by the judge, have the saddle taken off and to be led out parallel with the ropes and the onlookers for further inspection before being shown in hand by the rider, i.e. walked away on a straight line and trotted back past the judge the enable him to watch the pony's action.

He or possibly she will seek to choose the pony with the best movement at all paces when ridden, the best manners and standard of training, the most perfect conformation, quality and overall brilliance. But the pony's performance over fences and general ability and usefulness outside the ring will not come into the picture, although it is assumed that these will automatically be excellent if he has fulfilled the qualifications mentioned above.

He must be a true pony, not a beautifully made child's hunter, or small horse. Thus it is clear that in certain ways he is a type on his own, a model of perfection which at present makes the show pony classes very specialized and, perhaps, lacking in the diverse interest which characterizes so many of the overseas classes, and which must give far more pleasure and practical enjoyment to the young riders.

Many of our riders are as brilliant and delightful as the ponies they show, but not all can ever attain this standard, nor will their good but not exceptional animals gain many championships, although they compete regularly along with their more brilliant companions, and indeed fill and furnish the ring. Whereas if our classes included performance over fences and were as generally varied as those in Canada and America the children would not only get more pleasure out of competing in them, but would learn quite a lot of practical horsemanship which would stand them in good stead even if they did not gain the highest honours.

The good but not brilliant pony, properly trained, correctly ridden and in fit condition, can hunt and jump with the best, be capable of elementary dressage, have a fair turn of speed, perhaps even a certain amount of dash, and yet be a quiet and interesting ride for those who wish to jog around the countryside. They do all this very well in Canada – many are the perfect example of a good child's pony – so why not over here? Are we not becoming a little too 'precious', too ready to look for the picture pony, in our assessment of the show classes and in our efforts to breed the right type of animal for them, even to the point of overlooking some fundamentals and breeding more with an eye to the main chance, than of breeding, feeding and producing essentially good stock? This is an attitude of mind which needs to be guarded against, and which can have far-reaching results – bad ones!

In another chapter we will review to some extent the breeding of ponies in Canada and the States, the classes in which they are expected to compete and what they achieve. It suffices to say here that they must carry out work such as I have described previously, i.e. must hunt, jump, understand elementary dressage, etc., and take part in every type of performance class, as well as being shown 'on the line' – exhibited in-hand in various breed events, and also as models of their type – and all this in one or two day shows. They are hard fit, not just in show condition, and they think little of it. In some cases they are also driven, but this will be referred to later on, as at present we are considering only the type of pony we plan to breed, rather than the exact ruling of the classes – the ifs and buts of the show-ring events.

Each individual pony, and, of course, each breed, has its own natural gifts, in that it is better equipped physically and mentally and more capable of performing one job than another, but speaking generally and not individually there is little ponies cannot accomplish, both the mountain and moorland and the products of what is often a very expensive breeding policy. Sometimes there is a comparatively small amount of native blood in a sire who, if put to a good pony mare, will produce consistently good stock. In others the proportion is very large, and many registered native sires have produced consistently, sometimes startlingly, brilliant offspring. The right choice of parents and the right pre- as well as careful ante-natal care, plus a three-year happy 'childhood' for the resulting foal – are the vital requisites which cannot be ignored.

Too many people are breeding ponies without fully realizing

what they are doing, and why. This will not help the production of the good pony. It is most unwise to attempt to cash in on a rising market with indifferent stock. Breeding is bound to be a long-term policy and to attempt to bring into being a selection of animals which will fill a temporary need will certainly have a very adverse effect on the durability of that same market because, in the long run, it is always the pony of the right make and shape, whose parentage ensures stamina, a good temperament and a healthy constitution, who will be sought after. His less naturally, and hardily bred brother will not endure the test of time, nor will he have anything to hand down to his children and grandchildren. Therefore the breed or type will deteriorate and cease to be popular.

Does one divide the people who breed ponies into the following categories?

(1) Those who specialize in the apparently lovely sire or dam, so eye-catching that any slight weakness of hock, or any other limb or bone structure, or action is forgiven, hardly noticed or even accepted. Beautiful animals these, almost certainly show-ring champions in a small or possibly big way, yet not necessarily the progenitors of the versatile and handsome performer but rather of the type to which the eye of many modern observers has, alas, been trained. Their progeny may sell well all over the world, but will not add to the quality and standard of the next generation of ponies.

(2) The breeder who devotes himself exclusively to one breed, keeping it in natural surroundings which assist its development, with the right air, water, minerals and type of feeding, and who, above all allows it the freedom in which it can think, act and exercise itself.

This breeder has a real understanding of animals and a deep-seated apprehension and experience of what he is doing.

He inevitably produces very good ponies, often better than the youngsters bred from many a renowned show champion, and he never breeds indiscriminately. He prunes his stock with discretion, the less good ponies accepting a humbler walk in life which, in all probability, they grace exceedingly well. He has the seeing eye denied to so many of us.

(3) The man or woman who has gained his or her experience the hard but sure way, either by breeding and understanding one particular type of pony and then deciding that judicious crossing would produce an even more valuable article, and who sets about this experiment in cross-breeding with care, not expecting the first cross to be perfection but crossing again until the right pattern emerges. A study of the pedigrees of various ponies will help to make this point clear.

This breeder will probably produce some very good and valuable animals, using time as a friend, not an enemy, and not expecting immediate results.

(4) The individual who would like to breed one or two ponies as a hobby, or to have a foal from a beloved and ageing mare. If he or she will take the trouble to do this job properly they – and their ponies – may well have reason to be pleased, but if they treat it too much as an inexpensive hobby, which it is not, their unfortunate animals will suffer and so, probably, will their pockets.

(5) The avaricious breeder who has some knowledge and possibly plenty of downland available and who breeds rubbish to be bought by the unwise and the uninitiated, and who, alas, also sometimes sells his stuff abroad. Not only does he harm the breed by allowing indifferent stock to be purchased and bred from by the novice, thus weakening the overall standard, but he also harms the market by destroying confidence in the breed itself and in the integrity of those who sell it.

Leaving out the last paragraph, to which of the four other

categories do you belong? I hope that whatever may be your answer you will at least find something to interest you in this book and that the notes on our mountain and moorland ponies, which start on page 58, will help you to realize their potential and perhaps give you some ideas of how best to breed your ideal pony.

## Chapter 2

# HOW SOME FAMOUS PONIES WERE BRED

Have you an eye for the right type of sire and dam which will almost certainly produce the pony you want to breed? So much depends on this and on the faculty of refusing to be led away by flashy good looks as opposed to genuine worth, strength, stamina and overall balance. When breeding race-horses a careful study of the various blood lines involved, together with a certain flair, will help to put you right. But I would not say that this was necessarily the case with breeding ponies, although I agree that the dam, and her ancestry, are of great importance, just as they are in the breeding of Thoroughbreds. It is, however, not always easy to trace the dam's pedigree to any extent.

Some inexpensive mares have given birth to a string of offspring which have brought continuous success to their breeders. Take, for example, Gipsy Gold, dam of Mr A. Deptford's famous show pony, Pretty Polly, whose name will live for many generations in the pedigrees of her descend-ants – she may be said to have started a dynasty. After looking at Gipsy Gold carefully one could honestly say that her strong quarters, equally strong back (very important, this), fine limbs, pasterns sufficiently sloping but not too long, her depth of girth, shoulder, neck, charmingly proportioned head and generous eye made it difficult to fault her, and she looks as if she had the right temperament. In fact she was a picture of grace and efficiency and a credit to her Thoroughbred sire, Good Luck, and her dam, the Welsh mare, Tiger Lily. I am not precisely sure of her height, but it was, I think, no more than 13 h.h.

Gipsy Gold was mated with the Arabian stallion, Naseel, by Raftan out of Naxina, originally owned by the late Lady Yule (Hanstead Stud) and bought by Mr Christopher Nicholson for his wife, Mrs Nicholson of Kells, Co. Meath, Ireland. Mrs Nicholson also owned Gipsy Gold. Naseel was the right type of Arab to mate with this little mare, graceful, beautifully proportioned and an elegant, free mover. These two brought into being an impressive number of famous ponies for young people to ride before Gipsy Gold's death in 1954 at the age of 20 – all were good but Pretty Polly is the most famous. Perhaps this is not the most perfect example of the wisdom of sticking to a certain cross, i.e. if sire and dam 'click' and produce the right type of foal, go on pairing them, but it at least pinpoints a principle which is well worth remembering.

Mr R. S. Summerhays gave Gipsy Gold the following epitaph: 'In the case of Gipsy Gold we have something more than the end of an old favourite, for this mare has left behind her many ponies of great merit, some of which are not only outstanding but have even gone some way to establish a type of pony thought to be desirable in the pony classes.' His words are more than true, but what is so interesting is the success of subsequent generations of ponies which trace back to Naseel and Gipsy Gold and particularly to their daughter, Pretty Polly. They may differ widely in type, naturally depending on the outcrosses, but by and large they have maintained a very high standard.

There could hardly be a greater contrast than there is

between the two which I have chosen as an example, but both in their own way are proving the effectiveness of their breeding. One is Pollyanna, Pretty Polly's amazing daughter by Bwlch Valentino, of whom more anon. This lovely and invincible champion, 13·2 h.h. pony, full of quality and grace, achieved considerable honour in England and was sold to America for an astronomical sum, where she was re-trained to suit the American performance classes, in which she did very well before being put to stud.

The other is the more solid, part-bred Welsh 13·2 h.h. stallion, Mrs and Miss Stilgoe's chestnut, Cusop Policy, completely different in looks if not in character, who is by Bolgoed Automation out of Pretty Polly, and is siring some excellent children's ponies with quality and substance, thus carrying on the great tradition created by his dam. He lives a free and friendly life and is regularly ridden, all of which helps to keep him contented and fit.

Another little mare, this time of only guessed-at parentage, proved herself more than worthy to be included in any role of honour because of her looks, her character and the sons and daughters she bred. This was the much-beloved Welsh mare, Winkie, bought by Mrs Joan Smith-Maxwell as a just-weaned foal from some gipsies during the last war, for the sum of 50*s*. (£2·50). Her precise breeding is not known, but she was out of a mare owned by this gipsy family who was, to quote Mrs Smith-Maxwell, 'A great pattern and who had one eye, which my Romany friends insisted was great luck, and so it proved. As they travelled the Welsh side no doubt they had put her to a good Welsh stallion.' Fortunately Winkie is included in the Welsh Pony Foundation Stud Book, as her make and shape and way of going made it obvious that she could become a very valuable little brood mare.

This she did, having first assisted very considerably in the future riding career and general know-how of Ailsa Smith-Maxwell, then only seven years old, and now Mrs Nigel Pease. Her name appears in many illustrious pedigrees. Four of her offspring were Lemington May Day, Lemington Kismet, Lemington Gay Sprite and Lemington Carissima. Of these ponies probably May Day was the most famous, crowning her show-ring career with the supreme championship at the Ponies of Britain summer show in 1958. She was elected to appear at the Horse of the Year Show the same year, which, alas, was also the year of Winkie's death.

May Day was by the Arabian stallion, Rukuban, by Sahban by Alderbaran out of Seriya. Rukuban's dam was Roxana, by Ruskov out of Ranya. Roxana was a very famous mare who lived to a great age – Mrs Pease has a photograph of her being ridden by a child when she was 36 years old.

Gay Sprite was by the small Thoroughbred sire, Gay Presto, Kismet by Coed Coch Glyndwr, who was one of the greatest Welsh sires of his day and who has now gone to join the ranks of the immortals. And Carissima was by Phoebus, who sired so many good point-to-point horses in the Cheshire area.

May Day's continued success, first in the show-ring and now at stud, is certainly a feather in the cap of those who swear by the Arab–Welsh cross. Mrs Pease is a great admirer of this cross if children's ponies are the objective, as she says 'so often their temperament is extremely good'.

Here again much depends on the suitability of one horse to another – it is not always the most glamorous sire who will hand on to his progeny the qualities for which we look. How can he if his mate's proportions do not suit his own, however good either of them may in fact be? Further down the scale, and perhaps not entirely relevant to the foregoing remarks, if either the prospective sire or dam has any specific fault or weakness its mate should if possible compensate by being noticeably good in this particular point.

Just as the hand-picked native pony stallion crossed with a gracious and generously-made mare of his own breed will produce worthy sons and daughters, useful as riding ponies or for breeding, so will the good mountain and moorland pony, of either sex, if judiciously crossed with a Thoroughbred, Arab or part-breed, produce the type of pony likely to fetch a high price in the show-ring. But now, of course, we are coming into a far less simple and probably more costly branch of the business. How can one decide which stallions or mares are likely to give us the results we want? Presumably we have studied each breed of native pony and decided which is the most suited for our purpose and it remains to decide what cross we want to use.

It is easy to say 'I will cross my Dartmoor with a small Thoroughbred and something tells me that I shall get a very high quality little animal, likely to win a 13·2 h.h. class and certain to carry a beautiful pony head.'

But will you? Have you, for instance, been able to find the right type of small Thoroughbred likely to 'agree' with your little Dartmoor mare, to bring out her best points and eradicate those which are not as good . . . or even vice versa? Do you fully realize that although there are some excellent small Thoroughbred stallions available most of our valuable ponies are of a very mixed if highly illustrious parentage, as, for example, the inimitable Bwlch Valentino, owned by Mr V. Eckley, who has sired a large proportion of our most famous children's ponies.

Let us consider what he looks like, and his ancestry – the different stallions and mares who helped to make him what he is. A grey pony, standing 14·1 h.h. with that indefinable air of quality and the confidence given by a beautifully proportioned, highly efficient body and limbs that contribute to effortless movement, plus a noble head, and a kind, wise eye. He is a well-balanced pony, physically and mentally, and he has handed down these qualities to his many consistently good sons and daughters, filtered, of course, through their respective dams, the most outstanding of all probably being Pollyanna, whose grace and achievements, soundness and pleasant temperament we have already referred to.

This is Valentino and here is his family tree. He was bred by Mrs Nell Pennell and was by the polo pony stallion, Valentine, by Malice, who was by Maladante and was registered in the Argentine Stud Book. Did the Thoroughbred – polo pony combination help to implant not only the make and shape which gives an easy ride but also his long-striding action, sometimes described as 'floating'? He goes back through Le Sancy to Neopolis (G.S.B.). The dam of Malice was Alicia, by a polo pony stallion, Belsire, our of an Arab mare. His own dam, Bwlch Goldflake, was by the Thoroughbred, Meteoric, out of a part-Arab mare, Cigarette, whose dam, Sigorietti, was believed to be a Thoroughbred-Welsh cross who in her day won many flapping races for ponies under 13·2 h.h.

An interesting and well-documented pedigree. It would seem that each of these matings was correctly, if fortuitously, chosen and led to the production of a remarkable pony stallion, largely of Thoroughbred polo pony and Arab breeding with just a dash of good Welsh blood. This particular combination has created the substance, the lack of weediness, the easy movement and quality which are ideal in a child's pony. One of Valentino's sons, Mrs Colledge's Oakley Bubbling Stream, out of her pony mare, Bubbles, who is by the Arab, Count Dorsaz, has carried on the tradition of his sire by winning championships continuously since his earliest years, and here again we find a useful proportion of Arab-

Thoroughbred breeding if we study Bubbles' pedigree as well as that of Bwlch Valentino.

I could quote many other great stallions and brood mares, some of whom will go down in history as unique, and whose breeding is of equal interest, but having listed the various factors, the breeding, which helps to make a stallion what he is and the type of progeny he sires if mated with the right mare, it seems unnecessary to give further details of other famous animals of either sex.

There is, however, one very important point, which is fully agreed by most successful breeders of cross-bred ponies and which it is fatal to forget; always return at intervals to pure native stock. Only in this way will you retain the good pony characteristics. One might also add that the qualities of health, stamina, lack of hereditary disease (or any weakness of this description) inherent in these breeds, plus intelligence, enterprise and common sense, combine together to make a special, vital contribution. They are essential.

For your first cross never attempt to cross two native breeds – use an Arab or Thoroughbred, or a part-bred. At the next mating you can return to whatever breed you consider suitable, or have a yen for, remembering always to look for the stallion who will help to breed out his mate's faults or weaknesses, or any other unsuitability. But choose with discretion and an eye for a certain conformity, or you will produce a freak.

It is obvious that the only way to breed a pony of any height up to 14·2 h.h. is by the use of Arab, Thoroughbred and mountain and moorland blood in judicious proportions. The Arab's kind and generous temperament has always endeared him to the prospective breeder (but how important it is to choose only a good Arab); the Thoroughbred's high-strung courage, dash and quality, and the native pony's stamina, individuality and pony character (a definite quality but one which it is hard to define) all help to create that most endearing though often maddening animal of whose perfection we can at least dream.

## Chapter 3

# CHOOSING A MATE, MATING AND THE BIRTH OF A FOAL

It is not easy to advise on the choice of a stallion without knowing whether you are looking for a pure-bred mountain and moorland – and if so, what breed – a small Thoroughbred, an Arab, or a part-bred pony stallion with a good and interesting pedigree. And, most important of all, without knowing what is the end product you have in mind.

There are, however, a few general principles which are worth remembering:

(1) That the faults of sire and dam may well reappear in the progeny, even in an exaggerated form, so obviously parents both of whom have good average conformation are far more likely to produce consistently worthwhile stock than would be the case if either or both parents were outstanding in certain ways and bad or weak in others.

(2) Good temperament, balance, action and feet are all 'musts'.

(3) Choose a stallion who has the reputation of getting good stock. Do not rely only or necessarily on his looks (within reasonable limits). Study his pedigree.

(4) From the angle of your own business owning a stallion is a good proposition, as you can often save stallion fees and boxing to other studs. If he already has, or is likely to attain, a useful reputation as the 'getter' of the right type of stock, the

money he brings in will do much towards making your stud a financial success. He will help to pay your bills if you look after him well, that is, exercise, feed and ride him. Never let him be bored – this applies to all horses and ponies. More trouble and bad habits have started through boredom than many of us ever realize. Incidentally, if you own a stallion at stud you will require a licence from the Ministry of Agriculture and Fisheries.

Now for your mare who, we hope, has, like her mate-to-be, a good temperament, balance, action *and* feet. In addition to these invaluable attributes she should also have plenty of width across the hips and well-sprung ribs, plus a nice depth of girth. Strength over the loins is most important. Otherwise all you will expect of her is a conformation which is as good as possible overall, and which includes a sufficiency of width and strength behind her withers, where the saddle rests. I am not, of course, advocating too much flatness here or thick, coarse withers. We will assume that you know her breeding and have carefully considered it when choosing her mate.

Roughly speaking, mares come into use, or season, every three weeks during the spring or summer months until they are pregnant, but quite often they are irregular. They usually come into season (oestrus) about the ninth day after they have foaled, and some people consider this a good time to mate them, while others prefer to wait another three weeks. Should a mare come into use as early as the fifth day after foaling this is probably a false heat, and she should not be covered. It would be as well to wash her out with a mild disinfectant and not have her covered until the next time she is in season.

Never have your mares covered until they are really in use, for it is waste of time to do so when they are not ready. Mating is then against all their natural instincts and they will not conceive. To explain more fully: the ovum or egg which must be fertilized usually leaves the ovary of the mare and descends towards the womb before the spermatozoa from the semen of the stallion can reach it and fertilize it. The ovum only detaches itself when the mare is fully in use. Strangely enough this important factor is often overlooked and the resulting infertility wrongly attributed to the animals, although it is indeed the result of human ignorance or pigheadedness.

The ovum is shed from the ovary more towards the end rather than the beginning of the oestrus and there is no point in having the mare served on the first day, though possibly the second will do. In some cases they are served on the second, fourth and sixth. But there is no point (indeed it is not to be recommended) to have the mare served too frequently. If she is fully in use at the first service this may well be enough.

After she has been mated she ceases to be in season and if she is pregnant the next oestrus, due in about 16 to 18 days, does not take place, nor, in the normal way, will she be in season again until after she has foaled. But she should be carefully watched until ten weeks have passed by without her 'turning', and if possible it is a good thing to have her gently tried by the stallion.

To revert to the mating in the first place. If the mare tries to avoid the stallion's attentions it may be that she is not in use, or that she is nervous. Obviously it is waste of time to continue if she is not ready. In the case of shyness or nervousness, give her time to accustom herself to all that is happening and above all treat her with kindness and patience. To discover whether she is ready, lead her to one side of a stout gate or board, making sure there are no projections which might damage her or the stallion, and that she will not hurt her legs or trap them if she kicks. Matting on either side of a gate and a roller bar along the top may help in the case of larger animals.

The stallion then approaches the other side of the gate and they get acquainted by smelling each other. If the mare is fully in season she is likely to stand quietly and allow the stallion to nuzzle her, raising her tail and passing mucus and perhaps urine. If she is not ready she will probably lay her ears back, bite and kick at the stallion, or the gate. She may back away or refuse to come to the gate.

Stallions and mares running free on downland, perhaps in a herd, or at any rate in more natural surroundings than at many studs, mate at the most suitable time, and fertility is high. When a mare is at grass and a stallion is led beside the paddock rails she will call to him if she is in season, moving away from her companions and in his direction. Many stallions are very good with their mares, and the guardian, herd instinct is strong.

Some mares change their temperament when they are in season, a quiet one becoming more excitable and a fretful one more tranquil.

Your mare in foal should not be fat, but in healthy, normal condition. If you wish, she can be worked increasingly quietly until the last three months before she foals, when she is best kept at grass during the day. She must always have sufficient exercise or be free to exercise herself.

It is unwise to turn her out with geldings or in the next field to them. She might, in the early days, slip her foal as a result of the geldings' teasing her. In the more advanced stages she will keep very quiet if with another brood mare whereas geldings, who gallop about, might lead her into trouble. The same thing may happen in a hunting country, as the sound of the horn will cause mares who have been hunted to remember past thrills and gallop about. Better shut them up if hounds are coming your way. Over-excitement or pulling up too suddenly can cause abortion, or malpresentation of the foal at birth.

The danger of turning out ponies on ground which is thawing on top – i.e. greasy – but hard underneath is mentioned in another chapter and is particularly dangerous to brood mares.

She should have as much as she wants of good hay, with a feed of oats morning and evening. In the winter a bran mash mixed with oats and chaff three nights a week, also two or three carrots or an old mangold once a day will be of great benefit to her. Much, of course, depends on the season of the year and the quality of the pasture she is running in. It is important to keep her bowels open by giving her mashes as foaling time approaches, and for the previous fortnight a small dose of linseed oil should be added to her mash every other day, the amount required depending on the mare herself. Half a pint of linseed oil should be given to the average 15·2 h.h. mare (or smaller amounts according to the size of the pony) when foaling approaches and the wax drops off her teats.

Mares usually carry their foals for eleven months, or 340 days, but there are quite a few exceptions to this rule. As far as I know, the shortest time on record is 322 days and the longest 419. Watch, therefore, for any foaling signs about a month before you expect the foal to arrive. Her udder will increase in size, probably during the time she rests at night, although it may decrease while she is moving about during the day. Sometimes, but not always, a wax-like substance appears on the teats a few days before foaling. This may go in 24 to 48 hours and the milk appear.

When the muscles round her pelvis bones soften and drop to each side of her tail, and the vulva presents a swollen appearance, you may assume that the mare will foal soon, but during the day it is still best to turn her out and keep her under close observation.

It is not often that mares foal during the day – they prefer to do the job when all is quiet and they believe themselves alone. Therefore she must be watched at night, if possible without her being aware of it, or she will become nervous and delay foaling as much as she can. She should be observed or listened to every 20 minutes when she is near foaling – some people wisely make a little peephole from which they can observe her without being seen. If she stands quietly, or munches her hay, nothing is imminent, but if she is restless, walks round the box, paws the straw, lies down and gets up again, foaling is likely to start very soon.

The box should not be less than 14 ft by 14 ft, preferably 15 ft by 15 ft, and must be scrupulously clean – it should have been thoroughly disinfected. The straw bed should be nice and warm, but not too deep, and banked up against the walls and door to stop draughts. It is most important that the attendant who will cope with the mare and foal is scrupulously clean. A scrubbed bucket, filled with a non-irritating disinfectant dissolved in lukewarm water, a tablet of carbolic soap and a couple of clean towels should be available and arms and hands thoroughly washed before he or she goes into action. Sharp scissors, which have been boiled and are therefore sterile, a packet of sterilized navel tape, a bottle of iodine and sterilized cotton wool should all be available; also, possibly, an enema syringe of size used for humans, a bottle of liquid paraffin, a baby's feeding bottle with a large teat, Vaseline, a foaling drink for the mare and an ample supply of disinfectant. Some of these things may not be required but it is as well to have them handy.

The mare will be in need of nourishment after her exertions, so a warm bran mash with some oats in it should be kept in readiness. There must be easy access to a telephone, if this is possible, or some other arrangement by which a veterinary surgeon can be called without delay if the necessity arises. Normally birth takes place quickly, but if it is longer than usual it is as well to send immediately for a veterinary surgeon or it may be too late.

Any problems with a foaling mare are best dealt with under the direction of a man or woman who understands exactly what is taking place, and for this reason I strongly advise anyone planning to run a stud (*a*) to spend some while training and then working in someone else's stud (see final chapter); (*b*) to make quite sure that until they have gained this experience they do not attempt to look after a foaling mare without the assistance of a reliable and knowledgeable companion. No particular difficulties are necessarily involved, but the right 'know-how' is essential for both mare and foal. (An excellent little book on this subject and which is included in the Horsemaster Series is *Rearing a Foal*, by Jane Howell. Like this book it is, of course, published by Nelsons.)

After the navel cord has been cut the foal should be carefully dragged to the mother's head *if* she is not already on her feet, and she should be left to lick it and rest. The longer she lies and rests the better. If she is stabled, avoid wind and draughts and make sure she is not cold.

In many cases, particularly, of course, with the mountain and moorland breeds and if the foal is not too early in the year, they will benefit from foaling in a field. Let the mare have her bran mash, with a handful, or double handful, of oats in it (according to her size) and some chilled water. If in a box, clean it when she gets up, renew the straw but do not make her bed too deep or it will trip the foal when first it tries to get about.

Foals are usually on their feet, or trying to get up, fairly soon, but if after two hours it is still lying, get help. First put its forefeet out in front of it and gently assist it to rise. You and your assistant should stand one on each side, facing each other, each of you with one hand under the foal's neck, and the other behind its quarters.

Steady it and guide it towards its mother's udder, and, if it will not suck of its own accord, draw a little milk over your hands and fingers, let it savour and perhaps suck them. Then craftily guide its nose in the right direction – towards its mother's source of supply – and let it substitute her teat for your finger. If it is not frightened, patience and tact will soon teach it what to do. The mother's first milk is most important to the foal.

It is usually best to tie the navel cord in two places, and cut between them directly after birth. Cut about 2 inches from the foal's belly and tie with a piece of thin tape which has been soaked in disinfectant. The cut end of the stump should be cleansed with clean cotton wool soaked in disinfectant and then dusted with a little boracic powder, and this should be done twice a day until the navel is closed up. The mare should get rid of the afterbirth within an hour of foaling. Send for a veterinary surgeon if she has not done so within eight hours.

If the mare and foal are inside keep them there, very quietly, for the first 48 hours, making sure there is plenty of fresh air in the box. If the weather is suitable and the foal flourishing, lead her into a field and let her graze for half an hour, but keep her on the leading rein so that she cannot gallop about and injure or upset the foal. Remember that foals see very little at first, so ask a couple of friends to guide it to the field, holding it as described previously. After it has been out for a couple of times, one assistant to keep it from running down side tracks will suffice. Be careful that it does not dart between you and the mare and fall over the lead rein. And if it goes too far away, take her to it – do not attempt to drive towards her.

At the end of a few days they can graze loose but if the ground is wet bring them in should the foal lie down – if, in fact, it is tired. After a good rest in the dry box turn them out again for an hour or so. Gradually increase the time they are out until at last they are able to stay out night and day. If it becomes very hot and they have no shed or shelter bring them in at about 9.30 a.m. and do not put them out again until the cool of the evening. This gives you a chance to feed your foal on crushed oats and both animals will be protected from the sun and, up to a point, from the flies. The mare should have green fodder while she is in.

Remember that when the mare is in season (oestrus) the milk may make the foal scour. This is nothing to worry about, but make sure it does not catch a chill. Wash and Vaseline it under the tail to avoid scalding. Should it have acute diarrhoea, lessen the mare's food and water allowance for the time being, thus reducing her supply of milk, and let her have 1 oz. of bicarbonate of soda twice a day in her feed (less for smaller ponies), for preference in a mash, to correct the acidity of the milk. Administer a dose of castor oil to the foal to clear its inside of whatever may be causing the irritation.

Let your foal run at grass with his dam until well into the autumn, supplementing her food intake with bran and oats. When he is about three weeks old have him fitted with a headcollar which will let out as he grows. Slip it on gently. Two friends – good, strong ones – should hold the youngster, facing each other and each with a hand clasping him in front of his neck and the other round his thigh. Be particularly careful never to frighten him at this or any other time, or you will lay up many difficulties for the future. Also he may fall and hurt himself.

At the end of a day or two have the mare led round her box, attach a long rein to the foal's headcollar and gently persuade him to follow her. After a few such lessons take him to their paddock and continue the instruction, avoiding force as far as possible. The foal, must, however, realize that the rein is controlling it and that he must go where you wish. Two friends, one on each side, can help if he becomes obstreperous until such time as he realizes the foolishness of struggling. The lesson must only be a very short one. You must remember that he has only a baby mind and may take a long while to realize what you want, so never lose patience, expect too much of him, or condemn him if he does not respond.

Lead your foal each time he and his dam are brought back from the field, or as a daily habit if for any reason you do not need to bring them in. It is little trouble and will help him to learn. Handle him all over and teach him to have his feet picked up. Make sure they are in good shape and also watch his teeth. He should be weaned probably about mid-October, when he may be about six months old, but he must not be weaned until about four months after his birth, at the earliest.

Bring him and his mother into a box at nights and when he is accustomed to the box, starve the mare for 24 hours, with only a little water. The foal will then take all the milk she produces as the flow will have been checked. Persuade him to eat crushed oats and mash while he is in the box. After her 24 hours' starvation take the mare some distance away – where she and the foal cannot hear one another if they call, as they will, but leaving the foal in the box with which it is acquainted. It is a good idea to wean two foals simultaneously and then let them keep each other company in the box and afterwards.

If he has been having crushed oats and a little damped bran

since he was a few days old, and he is six or seven months when weaned he is not likely to deteriorate, or very little.

After weaning the mare must be watched. Feed only dry hay, and not too much of that for the first two or three days, and provide a short allowance of water. For the first day and night draw off the surplus milk every few hours, but never completely strip her or you will encourage the secretion of more milk. It will gradually become less and less, until she dries off completely. What you do will prevent the udder becoming distorted and swollen. If it becomes hard and painful, apply hot fomentations until it grows soft, draw off all the milk, rub the udder well with olive oil and administer a mild dose of physic.

The mare should be turned out to exercise herself, but not in a rich or abundant paddock or the milk will start again. Watch her for some days.

When your youngster has recovered from his weaning and is eating well, do not fail to dose him for worms. This is very important. Let him have plenty of liberty, if possible on ground grazed regularly by cattle, plenty of good food and protection from the sun and cold winds. The first two years are of the utmost importance and if you stint him of food and shelter, or if he leads an unhealthy life during this period he will never make up for it.

As one does not know whether you are planning to breed from a hardy mountain and moorland pony, a small Thoroughbred, an Arab, a part-bred, or a combination of two of these animals please remember that the advice given here is general and must be adapted to the type, breed and size of the pony – or horse – in your care. Quantities given may also need to be adjusted.

## Chapter 4

# THE CARE OF YOUR STOCK

The care of your ponies, whether they are of considerable financial value or not, should always follow a set pattern and must not be haphazard. This applies particularly to your brood mares and to your young, growing stock. Never believe that because a pony is of mountain and moorland breeding he or she needs little food and no looking after. Remember (*a*) that you have transported him from his native surroundings into far more limited ones; (*b*) that you want him to be fit to sell, show, breed from, or just do you credit – you do not want an under-nourished, ill-cared-for pony who may or may not have incurred the seeds of ill-health even if this is not immediately evident.

Warmth, food, freedom and companionship are perhaps the four most important, indeed, the most vital necessities of the pony at stud. Warmth means shelter in the winter and it results in a warm, healthy coat and feet in good condition, without which the pony will be at a constant disadvantage. Food must not only contribute towards keeping the pony warm, which is sometimes all it does, but must nourish and strengthen him as well. Freedom must enable him to exercise himself and to retain his individuality, his instinctive good sense and intelligence – in fact, to remain a free individual. It will also help him to keep his stamina and to breathe the fresh air which is essential for his blood stream and for his health generally.

And companionship, because without it he will not thrive as he should: he will not be contented and happy and he may even become mischievous and indeed a 'problem child'. The horse is a herd animal and it is not natural for him to be without companionship – of his own kind for preference, if not, of some other animal.

Your ponies are out at grass? What of the weather? How does it affect them? To sum up: if cold, frosty and sunny they will do well, but if wet, cold and muddy they will not. When looking for land on which to start your stud, whether of pure-bred mountain and moorlands or more highly evolved, cross-bred ponies, try to find well-drained land, part of it undulating and with water running through it or at least easily available. It should be well-hedged, or have trees which will afford protection from the sun and flies in summer and the bleak winds, wet and cold in the winter. If you have the right to use downland where there are warm hollows and copses and there is shelter from the north and east ponies will often thrive in these spacious surroundings, learning to come regularly and quickly for food when called. Rain and spring drinking water are excellent for them.

Brood mares will often do well if they are allowed to live out entirely (night and day) until Christmas, but after that they should be brought in at night and turned out during the day. This also applies to young stock, particularly foals and yearlings. Bring them in after Christmas and give them a warm, dry place to sleep. They will rest better, maintain condition and benefit more from their food.

When at grass make doubly sure that there is always plenty of fresh, clean water available, taking particular care that the ice is broken at regular intervals – in fact, whenever necessary

– on the water in the field or on the downs. Square boxes filled with sawdust can be used to encase the stand pipes of water troughs, and these will make them well-nigh frost-proof.

It is safe to turn out in frost and snow during the day, though not perhaps in a white frost which may cause colic, or if there is a thaw on the top of the ground which makes the going treacherous. Cold, driving rain and sleet will not help them – they are better sheltered.

At this time of year many studs feed about 15 lb of hay a day for mares of, shall we say, the size of the average riding horse, at grass and 10 lb for small mares and young stock, divided into morning and evening feeds. Also one feed a day each of 2–3 lb of crushed oats with added chaff and some sliced carrots. A feed of boiled oats with some well-cooked linseed is much appreciated if fed warm but not hot at night, especially in bad weather.

Mrs Colledge, breeder of many fine ponies including the already well-known Oakley Bubbling Spring, tells me that some of her mountain and moorland ponies stay out permanently as they have shelter and plenty of room to range. All are fed with as much hay as they want, in nets.

Her larger and more highly-bred ponies are in at night from December onwards and they also get two good feeds of say 4 lb crushed oats, 2 lb bran, a double handful of sugar beet, molassine meal, and a tablespoonful of cold liver oil, as well as Equivite or any other good mineral additive. Naturally these rations are adjusted to suit the various age groups and types. The brood mares are wormed four times a year with the utmost regularity, the remaining stock having Equizole worm pellets or powder, which, if they will not take it, can be given in a drench every six weeks.

Other studs are now using concentrates which are fed with

hay, sliced carrots, etc., and which are specially designed for mares and young stock. These are in the form of cubes and are widely advertised, the makers supplying full instructions as to the amount required and the method of feeding.

It is an acknowledged fact that a good, natural turf with a fair admixture of the vital, deep-rooted herbs is the best possible background for your horses and ponies. These herbs are chiefly chicory, burnett, narrow-leaved plantain and yarrow. Make sure you have them, if not introduce them. There should be more than 25 per cent clover and leguminous plants in grazing paddocks.

The paddocks must be well looked-after, grazed with cattle, topped, harrowed and rested, with a wise but not exaggerated balanced fertilizer treatment. Obtain sound advice from a good local farmer or from the Ministry of Agriculture and Fisheries on the care of your grass. And do not forget to pick up the droppings regularly when your mares are turned out. Not only is this most important because it benefits the paddock and prevents, or helps to prevent, worms, etc., but, in addition, horse manure will bring you in a nice little sum of money.

It is never advisable to keep your ponies in any one field for too long at a time, unless, of course, you go in for strip grazing with electric fencing, etc. Arrange for cattle or sheep to graze after them, or they can graze with the cattle, but it is best to let sheep in only after the ponies have been moved. Both help to keep the ground in good fettle.

If the grass has grown long you can first graze it with cattle, rest it for a couple of weeks and then let the ponies in.

Where there are good hedges and timber, especially oak, ash and thorn, ponies will thrive. Not only do these trees indicate that the land is suitable but they afford the essential shelter against cold winds and hot sun. These trees will not thrive on cold, wet land. If the field is fenced inspect the fence carefully for weak places.

## Chapter 5

# PRODUCING THE YOUNG PONY

In the previous chapter we studied among other things the necessity of feeding, worming and generally looking after a youngster properly, never forgetting tht the first two years of his life are very important ones. Incidentally, this care must be extended to his feet, a fact which, curiously enough, is often overlooked.

Some colts are of necessity gelded when a year old, but if possible it is best to let them reach two full years. Of course, this must depend on various factors, including whether your colt is to be turned out with mares and fillies. Do not, however, geld him until he is at least one full year. And think carefully before you make the decision to do so – you may be gelding a potential stallion of great merit! You must, however, cull your stock and always geld any but the top-grade ponies.

Under no circumstances break and work your youngster at two years old, although, of course one assumes that he has been handled, will wear a headcollar, is obedient to the lead rein, allows the farrier to inspect his feet and so on. On this subject I will quote the words of the late Captain T. L. Wright, M.V.O., M.R.C.V.S., which appeared in *Riding*. He says 'Fairly often advice is sought about a horse or pony which was "broken" at two years old, was a natural jumper, etc., and won many prizes at gymkhanas. Now at three or four years of age it has lost its action or speed or turned nappy.

'What can be done? The answer is that damage has been done during the growth that cannot be overcome. The only hope of salvage is rest for at least a year, and then an attempt to start, gradually, all over again. Neglect in this way is largely responsible for such defects as splints, spavins, ringbone and weak backs, or defective heart action. Quite recently a well-known racehorse trainer stated to the Press that "It is no good asking too much of a jumper if he is to last." '

Should you decide to exhibit your pony in in-hand classes, or, indeed, to compete in the ring at all, study carefully the rules of the breed society in which you are interested. For example, a Palomino of no matter what age must be shown with a full, unplaited mane and tail. A Welsh Mountain pony, or any other youthful representative of a mountain and moorland breed, is shown in all his own natural beauty, with flowing mane and tail, providing he is not competing in a ridden class when, of course, to plait and so forth is optional.

You can if you wish, trim off the long whiskers from the throat along the jaw line to the lower lip, also those around the nostril. When the edges of the ears are held together any protruding hairs from the inside can be trimmed off but do not cut the hair inside the ears under any circumstances – these keep the wet out.

A very good way of trimming the hairs of the cheek and jaw line is to start at the top and work down, as a hairdresser would, lifting a layer of hair through the mane comb and snipping the protruding ends. If the tail is very untidy a few of the short hairs underneath the others can be pulled to help make the remainder lie flat. Or the top of the tail can be plaited. Remember that when your youngster progresses to the riding pony breeding classes and his mane and tail are plaited his heels must be trimmed. If the mane is not plaited the heels should be left alone, except in the case of an overabundance of feather, when a little thinning (not too much) with comb and scissors is admissible.

As well as studying the rules of the society under whose auspices you are showing, and the wording in the schedule of the class in which you propose to compete, it is always a good thing to try to understand what is wanted in the various in-hand events. You will learn more this way than any other. If at a show something seems contradictory, or you do not understand it, inquire from any breed representative present, or from some knowledgeable person including the judge himself after he has finished his job. A good judge will often teach a genuine novice a lot.

Do not under any circumstances over-show your young stock even if you are tempted to do so because (*a*) you were so successful; (*b*) you did not win but were passably well-placed and thought you would do better next time. An occasional jaunt to a show will not hurt a young pony, too much of it most certainly will.

If you decide that once he is of an age to be broken and ridden you will exhibit your handsome novice in a class for children's show ponies you must plan his career wisely. One of our most consistently successful exhibitors arranges to have his ponies broken quietly, correctly and without hurry when they are three years of age, but they are never expected to work seriously at this age, although they may be taken to one or two shows towards the end of the season in question, just to see how they react when ridden in the ring and to let them get the feel of it.

The following season they are shown more seriously, and probably gravitate quite quickly from the novice to the open classes, possibly winning both at one show. But they do not attend too many shows, do not become bored, nappy, ring-shy, frustrated, unsound, or anything else, and are not shown at any fixture during some special holiday period, when there is likely to be too much traffic on the road, or too much hanging about in the box or out of it.

As this exhibitor's ponies are never bored, yet never produced too quickly (which could puzzle them and cause them to set up a resistance) and live a normal pony's life (at grass, hacking, hunting, etc.) their temperament is usually unspoilt and they are schooled and ready to make a career in any equine walk of life. They look well, go well – and sell well!

Recently I glanced at some photographs of our old, retired show ponies in the *National Pony Society Review*, and the history of contented service and friendship was revealed by the captions. Royal Show, who, in his prime, beat Pretty Polly towards the end of her career, has spent his latter years quietly hacking and hunting with the Cotley Harriers – a noble pony this, who seemed to thoroughly enjoy his time in the show ring charmingly partnered by Jenny Bullen – now Mrs Loriston-Clarke. Silver Moon, photographed at 21 years carrying Jenny's daughter, Anne, then aged 21 months. No one held Silver Moon, or Anne, as these two had been friends since the latter could sit up in her pram. He had a remarkable history as he came out of a knacker's cart in Ireland, too weak at first to be shipped to England. Three months later he was first novice and second open 13·2 h.h. prizewinner at the Richmond Royal Horse Show.

The photographs also included the golden Palomino stallion, Bubbly, who belonged first to her parents, Lieutenant-Colonel and Mrs Bullen and is now owned by Jenny Loriston-Clarke. I have seen this stallion carrying children bareback outside the show ring. He is a great lover of human beings as a whole, especially children, has won innumerable championships, is still siring wonderful stock and is still ridden every day. At the time of writing he is 22.

The beautiful Connemara, Miss Ginette Mason's Conquest of Calla, who won his first novice one-day event at the age of 18 – a fine dressage, cross-country and show-jumping performer, as well as a good pony to hounds. He loves his outings, although he was officially retired from the show ring at 16.

So the list goes on, and one can think of many more, sufficient indeed to fill a book. These old show ponies and a great number of others like them are in many cases breeding first-class young stock. They were never looked upon as expendable by their owners, or pushed casually on one side when their show-ring career seemed to be over. They are all friends of the family, kind, generous, highly individual and with a great interest in life all around them. When the time comes for them to go to another world they will be deeply missed and leave behind them a memorable record of service, willingly rendered.

Chapter 6

# SOME FIRST PRINCIPLES

A full programme of instruction on the breaking and making of a pony would be out of place in this book but here are a few first principles, some of which are equally applicable to the mountain and moorland and the child's show pony who can no doubt trace his ancestry back to one of these invaluable native breeds.

Your pony must be kindly but firmly educated, not 'broken'. Much that you will have to teach him will seem to him to be alien and pointless until he either understands what it is all about, or has profited from it by becoming supple, handy and obedient. Therefore you must be patient and remember that 'slowly, slowly' are your watchwords. They will show you the way to success but only if you accept and understand your pony's limitation.

When asked 'What makes a show pony?' the late Mrs Anne Bullen, who herself understood so well the ponies, their riders and how to produce both, said: 'Is it not the same thing that makes a "show girl" – good looks and hard work!' How right she was, but the first steps taken by the young pony in this direction must be easy, not hard. Later he and his owner (especially the owner!) must really work – and, we hope, enjoy their work. Mrs Bullen adds 'for either the girl or the pony to rise above the average they must have an extra gift.

Glamour or personality, call it what you will, it is this extra gift that will make either the "top show girl" or the "London pony".'

Two things occur to me in this connection: One, that given time your top class pony will develop a sort of pride, an air which will make him enter the ring with great assurance and which will inevitably cause him to become the focus of attention. This is 'presence' and will naturally take a little time to achieve.

The second is that his brothers and sisters, very good in their own way but lacking this superlative brilliance, are equally capable of carving a highly successful career in performance classes, gymkhana games, Pony Club events, hunting, hacking and so on, and they, too, should be correctly educated and able to behave correctly in the show ring.

Alas, too often ponies are given no chance to learn their job – they are taught nothing properly, incorrectly looked after and unreasonably expected to know better than their owners – which, in fact, some eventually do! To muddle and perplex a pony, or make a fool of him, will often cause him to turn awkward or nasty, whereas with correct schooling and the right sort of life it is amazing what he will be able to accomplish.

Here are a few hints on the production of a pony who is to compete in the ridden classes:

(1) He must have systematic work, not prolonged to the point of boredom. This to include work on the lunge until he carries himself well and has a certain rhythm in his movements – in his way of going.

(2) Proper condition, which will build up from feeding, worming and right exercise.

(3) Lessons in obeying the aids when ridden. Too much turning and circling should not be asked of him and his

schooling must be relieved by pleasant hacks across country, which are in themselves educational. They teach the pony to use himself at all paces and in different circumstances; they also teach him a happy combination of obedience, initiative and self-confidence and provide a variety. Leave his head free on these excursions.

(4) Make sure that your pony's head is carried correctly when you are schooling him for the ring, never relying on a double bridle to get his nose in a good position. If he has learned to use his hocks and his rider's hands are sympathetic this should prove no great problem.

(5) See that his manners are good and that he has confidence in his rider.

(6) Trim his jaw, ears and heels, pull his mane and tail. The tail may be washed but neither a dandy brush nor a comb should be used, as they will ruin it. Brush from the top with a body brush, but do not brush out the ends or put on too tight a tail bandage, or leave that bandage on all night. The tail can be damped if necessary but not the bandage.

(7) Light shoes or aluminium plates without clips help to improve a pony's action in the ring.

(8) Never fail to get to a show in good time, allow your pony to become accustomed to the surroundings and, if advisable, give him a good 'work-out' before he goes into the ring.

(9) Choose, and take care of good tack. Without knowing your pony it is difficult to advise what is the right bit, but in all probability a short-cheeked double, not thin or sharp, is the right combination for the open classes. A short-cheek Banbury which has a moveable mouthpiece might, also, fit the bill. The bit must be the correct width for the pony's mouth and the bridle must fit comfortably.

Obviously the fit of the saddle is equally important and it is an excellent idea to have it made to fit the pony. If you cannot do this, take your pony to one of the bigger shows and when all is quiet lead him to a saddler's stand. The saddler will no doubt let you try a few new and second-hand saddles on your pony and give you some useful advice. Or if you get in touch with him before the show he will send some saddles of a suitable size to your horse box near the ring for you to make your choice.

We will finish this chapter with two very simple tips. If you are teaching your pony to accept his snaffle bit early in his training, and all he wants to do is to spit it out, in spite of the offer of tempting morsels of apple or carrot to munch, offered in the hope that these will start him playing with the bit, then try the following.

Pull some long hair from the mane or tail and make it into a fairly thick tassel about an inch in length. This should be tied to the centre joint of the snaffle bit and adjusted so that it almost touches his teeth. If this is not enough to make him play with his bit for a while then smear the tassel with black treacle and he will suck it and forget his sulks.

Later on, when he is ready to compete in the ring you may wish to increase his glamour with a pattern of diamonds or squares on his quarters. These water marks can only be achieved successfully on a fine, soft coat, and should cover the flat part at the top of the quarters on either side of the spine. Use a damp water brush to brush the coat in different directions, and start by brushing it on top of the quarters towards the rear – in the direction the hair grows.

Now place the point of the brush against the spine and draw it down in a straight line towards the ground. Fill in the space with these strokes, each time placing them the width of the brush apart. Using the brush in the same manner draw equally spaced lines from the loins to the rear. This will complete a series of squares the width of the brush in size.

All lines running in the same direction must be parallel or the pattern will not be symmetrical. Do not make the brush too wet or the hair will be crinkly when dry. After finishing your pattern brush the hair below it downwards to a line level with the stifle, then horizontally down to the hock. This will emphasize the width of the lower part of the quarter. These squares or diamond marks are illustrated in the photographs of Bennochy Ailsa Craig and Pollyanna between pages 56 and 57 of this book.

It takes a real craftsman and a lot of practice to get the hang of making a nice chessboard pattern of water marks, but it is not difficult once you have really worked at it. Why not get a large piece of velvet, stretch it tightly across a board then practise on it with a water brush. At least it will stand still and not get bored!

## Chapter 7

# RULES FOR SHOWING PONIES

Wise rules in connection with the showing of ponies have been made by the various societies connected with them, such as those included in the annual handbook of the British Show Pony Society, and the Rule Book of the Canadian Horse Show Association, Inc., and others. These are constantly revised to bring them up to date and are available each year. They are very important, are framed to help and promote the correct showing of ponies and without them there would be disaster in the show-world.

What is so amazing is that many exhibitors do not study them, to their own undoing and I suggest that it would be profitable to obtain copies as soon as they are printed and are available from the relevant society. This particularly applies to the British Show Pony Society.

The rules of this society, which are listed in the annual handbook, deal with a number of vital points, such as the age at which ponies may be shown in the ridden classes. At the time of writing three-year-olds may be exhibited in novice events only on and after 1 July, must wear snaffle bridles and may not compete in championships.

With regard to the ruling on unsoundness I will quote the relevant pargraph in the 1970 B.S.P.S. handbook. As careful thought was given to the wording of this rule it seems unlikely that it will be changed quickly. It reads: *If in the opinion of the judges a pony is considered to be unsound, the exhibitor should be given the option of withdrawing the exhibit from the class or asking that the official veterinary surgeon should examine the pony. The veterinary surgeon's decision is final.*

When judging in Canada I studied the Canadian Rule Book on this point. It states: *Each recognized show must have a qualified veterinarian present whose decision as to soundness of the animal shall be final.* (*a*) *If the veterinarian is not immediately available, the exhibitor's statement as to unsoundness shall be accepted for the purpose of excusing the horse from showing;* (*b*) *If a veterinarian is not immediately available the judge's decision is final in respect to a horse in competiton.*

*The examination of a horse in the ring by a veterinarian shall be done as inconspicuously as possible and in such a manner as not to draw public attention thereto. The co-operation of judges to this end is required.*

With regard to the last paragraph from the Rule Book, although the problem did not arise in the classes I was judging I discussed it with one or two people and learned that in Canada the veterinary surgeon will, in the case of possible lameness, come into the ring when asked for and will stand beside the judge when all the ponies are going round. If one of them is lame he will himself pick it out. I assume that otherwise the class is allowed to continue without the implication of possible unsoundness in any particular pony.

To continue with the rules of the British Show Pony Society, which cover subjects such as objections, use of stimulants, spurs, height, shoes, leading of ponies, galloping and many other points, the following statement is of importance: *No person may deliberately enter and show a pony under a judge who is known to have bred, sold or produced that pony.* (Panel judges are asked to take particular note of that rule).

The handbook also contains a page devoted to the rules for judging working hunter ponies. Here are some extracts, which will give you an idea of what is wanted:

AGE OF PONY:
four years or over.

AGE OF RIDER:
13 h.h. 14 years (rider not to have attained 14th birthday by 1 January).

14 h.h. 16 years (not to have attained 16th birthday by 1 January).

15 h.h. 18 years (not to have attained 18th birthday by 1 January).

HEIGHTS OF FENCES:
13 h.h. minimum 2 ft 6 ins maximum 3 ft.
14 h.h. minimum 2 ft 9 ins maximum 3 ft 3 ins.
15 h.h. minimum 3 ft maximum 3 ft 6 ins.

SADDLERY:
Snaffles permitted. No change of saddlery at any phase. No spurs.

MARKS:
No time limit. Jumping, 50; type, conformation and freedom of action, 30; style and manners, 20; total 100.

PENALTIES:
First refusal 10, second refusal 20, third refusal disqualification. Knock-down 10. Fall of horse or rider 20. (A complete turn-round constitutes a refusal.)

Any horse or pony that has won £25 in prize money under B.S.J.A. rules up to the closing date of entry will not be eligible.

There is quite a large number of classes for conformation hunter ponies under saddle in Canada. The jumps vary in height according to the size of the pony, as they do in England, but the judging is 60 per cent for performance, manners and way of going, 40 per cent for conformation, quality, substance and soundness.

Rule 27 of the Canadian Rule Book may appeal to those of our ringside connoisseurs who say that our in-hand ponies are often better turned-out than some of the people who lead them. It states, very politely, that 'It is the tradition of the show ring that an exhibitor be correctly dressed for the class in question, that attendants be neatly dressed, and horses properly presented. The management may in its discretion bar any entry or person entering the ring if not suitably presented to appear before an audience.'

In the same book the pages devoted to the judging of Welsh classes give some idea of the esteem in which this pony is held. There are events for Welsh trail ponies, in English or Western equipment, pleasure ponies, driving, including roadster, and many more. The qualifying gaits are stated to be (*a*) *Walk: true, fast, flat-footed, elastic and showy.* (b) *Trot: square, collected and reaching.* (*c*) *Canter: smooth, slow, collected and straight on both leads. Emphasis on all gaits placed on boldness; speed penalized except on Roadsters.*

These notes on different types of class and the various rulings may seem irrelevant in a book of this description, but are included as an indication to breeders of what may be expected of their ponies, to give them an idea of how they must be trained and what they must learn to understand if they intend to go in for showing to any great extent. There are, however, plenty of simple and straightforward classes in which they can gain experience without undue difficulty, if they so wish.

The placings often seem puzzling to those standing on the ringside, particularly in the in-hand classes. Occasionally the critics may be right, but it is as well to remember (*a*) that one cannot see from a distance what the judge can observe and study when he is close to the pony, or indeed, when the pony is run on a straight line towards him. Your view from the stands is not from directly in front or directly behind the pony and his action may be less true than you imagine.

(*b*) A judge must pay attention to movement, to type, if in a mountain and moorland class and to suitability for the job the pony will probably be expected to do; he must notice if it has plenty of heart room, is well coupled, with the tail nicely set on, that the back is not dipped – or roach, for that matter – and that the pony is well turned out and well presented by the handler, always bearing in mind that this is, after all, a *show* class and the best must win. It is not always realized that an animal with good conformation and fine proportions often looks smaller than actually he is.

(*c*) Before blaming the judge for what you think is a mistake read carefully the wording of the schedule or programme – the winner must fulfil these specifications.

(*d*) If there is little to choose between those at the top of the line, remember that each judge has an individual taste, a slight leaning towards one type more than another. He may go a little more strongly for certain points in conformation, or attribute particular importance to movement. He may prefer the Thoroughbred type of pony or the one with slightly more substance.

Judging is not easy, especially when it comes to deciding

between two or even three ponies of almost equal merit. It is even more difficult in a mixed class in which there is a disparity in age and sex, or in a mixed mountain and moorland class when the breeds compete against each other – here one must find those ponies which are indeed the best specimens of their respective breeds and it always gives an appearance of favouritism if the judge is obliged to put up two of one breed simply because those two are better than anything else in the ring; or a supreme championship when several animals of different type, age, sex and value are fighting it out.

The late Mr Horace Smith of the Cadogan Riding School who had a very considerable experience of horses, showing and judging, always insisted that there should be a certain uniformity of type in the line-up, that those at the top should all conform as far as possible to a certain pattern.

It is interesting to follow if you can the judge's way of handling a class, to understand his reasoning and placing even if you do not agree with it. If you can do so this does at least show that mentally he has a certain pattern in mind and is working to a set purpose. I have always been well pleased when breeders, exhibitors or other observers have told me that they could see what I was looking for when judging a class or a championship. It does at least make it more interesting for the onlookers.

Speaking of type – if you wish to win, breed always for type and suitability for the career for which your pony is destined. Our native ponies can be improved by selective breeding but great care must be taken to keep them true to type and to maintain the best points of the breed they represent. This applies to all ponies.

One other word about judges. They are human, they can have off days and make mistakes, but it is generally agreed that their greatest watchword is integrity. If a judge genuinely considers that the pony he puts at the top should win the class his opinion must be respected.

If you want to help the judges and yourself into the bargain, pay heed to the following very elementary tips, which, strangely enough, are often disregarded.

(1) Pass, yes, but never crowed upon, or stay beside another pony in the ring, whether led or ridden, because the judge will get a far better view of your own pony and how it moves if this simple advice is carried out.

(2) Ask yourself why the judge wants you to run the pony in hand. If you come up with the right answer (i.e. because he wants to see if the pony moves straight and freely, does not go too close or too wide whether in front or behind, does not plait or dish, and has a nice length of stride) then you will make it as easy as you can for the judge to get the best possible view. You will lead your pony out of his place in the line-up and make him stand straight and square between the judge and the ringside, so that the judge has a full view on one side and the spectators on the other. The judge will look carefully at the pony – if it has been competing in the ridden class this gives him a chance to see it without its saddle – weighing good points against less good ones, and feeling the limbs and possibly the feet for weakness or blemish.

At the word of command you will walk in a *straight* line away from the judge, *not too far*, turn with your pony's head on the inside – you should be on the outside of the pony on the turn as you can control it better this way – and trot back at a steady pace, yourself to one side and slightly ahead of the pony, still on a straight line, *not* at an angle of 45 degrees, and pass the judge, before slowing up, to enable him to watch the action of the hind legs. Do not hold the rein too close to the pony's head.

You will then lead it back to your place in the line-up,

taking every care not to pass too close to the hindquarters of the other ponies. All this applies equally to the run-down in-hand of the ridden pony classes and the ordinary in-hand events.

Beyond possibly a polite 'good day' it is best not to speak to the judge unless you have very good reason for doing so, or in answer to a question.

Remember your pony's good points and give these due significance in your method of preparing, producing and handling him. You may feel a little depressed if there are many ponies of outstanding merit in your class – at home your ewe lamb seemed almost without equal, now you fear that he is outclassed. But if you keep up your spirits and his, and show him to the best possible advantage you may find that his temperament, gaiety, manners and overall merit may compensate for his lack of brilliance and bring him a rosette.

Those of us who love our native breeds, fully described in the remaining part of this book, may like to know that in Canada Exmoor ponies are much respected for their temperament and their way of going in performance classes over jumps, both in the ring and across country. These they often win, or else are 'in the money'. They are also successfully used for breeding and for crossing with other stock. Welsh Section B ponies compete extremely well in these classes, although sometimes they are a little high-couraged. They also do well in driving events.

Now that Western classes are becoming popular in England we might well include in our programmes events such as the following:

Saddle type pony, Western equipment, open, 12·2 h.h. and under, or over 12·2 h.h. and not exceeding 14·2 h.h. Natural mane and tail. To be ridden by a Junior Exhibitor and show at a walk, trot and canter both ways of the ring. To be judged 60 per cent on performance and manners and 40 per cent on conformation and soundness. (Appointments: plain stock saddle. Curb chains optional. Rider to be attired in Western type of clothing and boots. Ponies may be any colour or breed.) Junior Western Pleasure Horses. Open to horses and ponies to be shown by a Junior Exhibitor. To be judged on performance 60 per cent, conformation 30 per cent and appointments 10 per cent.

In the Western Pleasure Horse Section referred to in the Canadian Rule Book the following rules for judging are outlined.

(*a*) Horses to be shown at a flat-footed walk, jog, lope and hand gallop both ways of the ring on a reasonably loose rein and on the correct lead at all times. Extreme speed should be penalized but horses should exhibit reasonable speed and be well under control of rider at all times. Horses should have reasonably good quick stop and should stand quietly after loping or hand galloping. Horses should back readily with minimum pulling on reins. Riders may be asked to dismount and mount while horse stands quietly.

(*b*) Tack: Horses to be shown under Western saddle and any humane type Western bit of hackamore. No running martingale or tie-downs permitted.

(*c*) Judging: Performance 60 per cent; conformation 30 per cent; appointments 10 per cent.

At the Arab Show at Barrie, Ontario, recently, young people also competed in the half-bred Arabian Pleasure Horse (Western) class, shown at a flat-footed walk, jog-trot, lope and hand gallop, with the emphasis on a loose rein at all paces, no martingale or tie-down. These classes were 'worked both ways of the ring and otherwise at the judge's discretion.' Some of these half-bred Arabians were ponies.

There is, in fact, such a vast selection of jobs which a pony can do, and classes in which he can compete, either in-hand, ridden or driven, that I think we may look upon him as one of our most valuable assets as well as a beloved friend and companion.

In the foregoing pages I have tried to show how breeders appreciate our native stock and how, in some cases, they have made use of it. The following details of the origin, type and qualifications of these mountain and moorland ponies will, I hope, interest you. How they have been used in successful crosses has been described in certain instances and some of their most famous descendants referred to. This may help you to realize their value and also to decide on the best possible way of utilizing them for stud work. The breeds appear in alphabetical order.

Chapter 8

# CONNEMARAS: HOW THEY EVOLVED

The origin of the Connemara pony has been widely discussed and many inconclusive if dramatic or romantic reasons given for its existence. That it is hardy, good-looking and has an agreeable temperament is indisputable, and it is quite possible that at different times Arab sires have added their valuable quota of good qualities to its make-up. The theory that these sires were shipwrecked and swam ashore is less probable than the more practical suggestion that the merchants from Galway who traded with Spain brought back or imported the odd Arabian stallion. It is said that many Andalusian horses were wrecked on the coast and joined forces with the Connemaras perhaps 300 years ago, and, possibly later, a little Breton, then Clydesdale blood was incorporated.

It is thought by some people that the introduction of Arab blood may have led to the appearance of blue-eyed, cream ponies which are not accepted for registration by the Irish Connemara Society. They maintain that the mating of two light-coloured ponies of Arabian ancestry will result in this defect, spoken of as 'a weak-eyed strain', and one which they wish to eliminate, although they are still accepted by the English Connemara Society. Many of the Irish farmers who are interested in cross-breeding send their mares to a Thoroughbred rather than an Arabian stallion.

The Celtic ponies, from which the modern Connemaras stem to a large extent, arrived on the Western coast of Ireland 2,000 years ago, fleeing from the Roman legions. Swift, strong, dun-coloured ponies these, and no doubt glad to settle in so well-watered and spacious an environment as Connemara, which they reached in due course. Here frequent changes of pasture were easily available. They mated with the ponies indigenous to the area and were little noticed by the outside world for about 1,600 years.

That the breed profited from its environment cannot be disputed. Phosphates, limestone and a climate made it possible for good herbage to be available with little loss of quality from early spring, right through the summer and even much of the winter. I believe that some of the sedges start to grow early in October and gorse and heather are also eaten by the ponies before spring breaks through. Add to this the hills, moors and valleys which, in addition to providing food and shelter, encourage the ponies to 'use themselves' to the fullest extent, developing their limbs, sturdy conformation and initiative, and it will be seen how and why the ponies grew and developed as they did.

Five distinct types of Connemara pony contribute to the pony of today, the fifth of which may be considered as the Clifden type, which is probably the most direct descendant of the Celtic pony, whose origins are so far removed in time that it is difficult if not impossible to trace them to their source.

Thus the Connemara pony is immensely interesting, and it is not surprising that he is much valued for his own achievements as well as for cross-breeding to produce horses of the calibre of the show-jumper Dundrum, ridden with such success by Tommy Wade and champion in Ireland, England, Brussels, Amsterdam and elsewhere. He won the King George V Gold Cup at the Royal International Horse Show at the White City and was a member of the team which was awarded the Aga Khan Trophy (Nations Cup) for Ireland. Dundrum was out of the registered Connemara mare, Evergood, by a Thoroughbred stallion, Little Heaven. He was bought as a yearling by Mr James Wade in County Tipperary.

A Connemara mare was the dam of Korbous, who gave such a pleasing exhibition of dressage without a bridle under the direction of Miss Penny Moreton at the Horse of the Year Show at Wembley. His display included jumping without wings. Korbous was versatile to a degree, with a wonderful temperament, heaving been hunted regularly by Miss Moreton and proving himself to be a talented show-jumper into the bargain. As the horse, Cooper Coin she had arranged to ride in the Three-Day Event at Harewood in Yorkshire in 1957, was sold to America, Penelope Moreton rode Korbous at this event and this remarkable combination came second to Sheila Willcox on High and Mighty.

Korbous continued to give a very good account of himself both as a jumper and a dressage horse, but rather than list his successes it might be more interesting to speak of his romantic parentage. His sire was a North African Barb, Carlore, who arrived as an unbroken four-year-old with a batch of mules from Morocco to the 4th Pack Group, R.S.S.C., commanded during World War II by the late Colonel Joe Hume Dudgeon. Their duty was to carry supplies to the troops in the mountains. Carlore was allotted to Stewart Murless, the brother of Noel Murless and himself a racehorse trainer in the Curragh and these two saw service in Tunisia, Sicily and Italy before Stewart brought Carlore back to the Curragh.

Eventually he was mated with a Connemara mare belonging to Mrs Brown of Bagnalstown, Co. Carlow, and when the resulting offspring reached two years he was offered to

Colonel Dudgeon as a present. The Colonel said he'd love to have him but sent a token payment in the form of a cheque to Mrs Brown (which I believe was returned) and called him Karbous, after the village near Tunis where he was bivouacked when Carlore arrived with the mules. He broke him at three and then sent him back to grass for two years to develop and grow up.

Another very interesting result of crossing Connemara and Thoroughbred was the elegant and highly receptive Little Model, who was by Dundrum's sire, Little Heaven out of a Connemara mare. The late Mrs V. D. S. Williams trained and rode Little Model to win national and international dressage competitions. They were part of the British team which won the European Dressage Championship in Copenhagen, when Mrs Williams was awarded the Carven Cup for the best lady rider.

Many other pure-bred and part-bred Connemara ponies have proved how useful they are by carrying children and adults with safety and success and they are popular in many countries, including America and Canada where they have their own stud books.

Their height usually ranges from 13 to 14 h.h., although another inch or two will not debar them from the English Connemara Pony Society, of which Mrs Barthorp is the secretary. Her address is The Quinta, Bentley, Farnham, Surry. A breed show is held annually. The accepted colours are grey, black, bay, brown and dun, and there are also occasional roans and chestnuts. They should not be long in the back and their bodies should be deep, with sloping shoulders and a graceful head and neck. They should be generously proportioned, capable of carrying a surprising amount of weight, good, light movers and have a graceful head and neck.

The secretary of the Connemara Pony Breeders Society in

The Welsh Pony of Cob Type. Mrs J. Crotty's Turkdean Cerdin, a
ır-year-old chestnut stallion by Llanarth Firel out of Revel Romance.
cond to Llanarth Meteor at the National Pony Show at Stoneleigh,
69.

A powerful Welsh cob with a noble outlook, Tyhen Comet, owned by
r J. Hughes of Cardiganshire and bred by Mr W. H. Davies. He is by
ntre Eiddwen Comet out of Tyhen Mattie. At Wembley 1969.

3 Rudi, Mrs L. Ensten's thoroughbred pony stallion, supreme champion pony stallion at the Ponies of Britain Stallion Show in 1968 for the second year in succession.

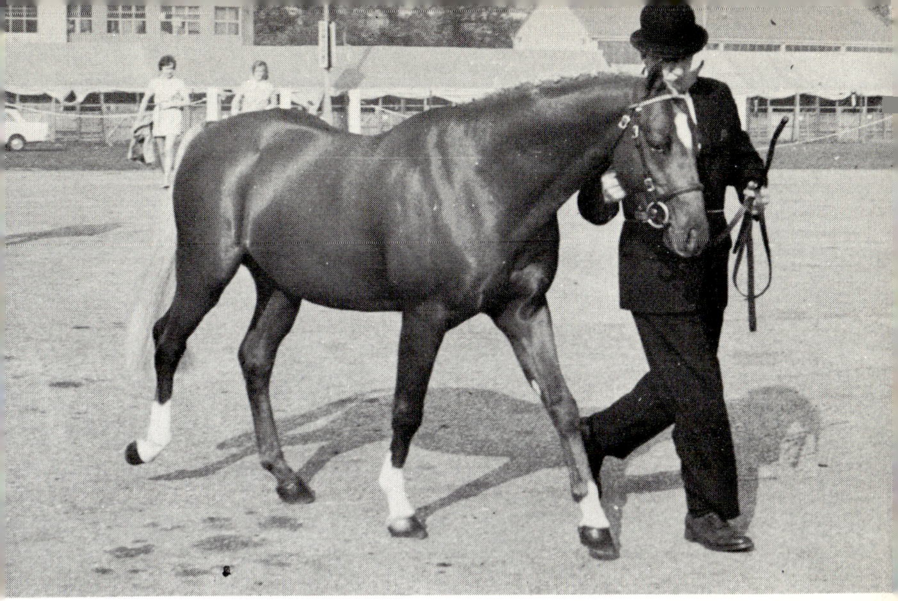

A very high-quality pony stallion, Oakley Bubbling Spring at three ars old. He is owned by Mrs S. Colledge and is by Bwlch Valentino out his owner's old show pony, Bubbles who in turn was by the Arabian allion, Count Dorsaz out of another well-known show pony, Catriona. akley Bubbling Spring has won many championships.

Mrs Dorian Williams's beautiful stallion, Pendley Model, at three years d, when he stood 14.1 h.h. This chestnut is by Solway Master Bronze it of Pendley Lady Precious Stream. Solway Master Bronze was by the ng-famous Coed Coch Glyndwr out of the Welsh pony mare, Criban ddy Bronze. He has had many triumphs in the show ring.

6 Supreme mountain and moorland champion and winner Welsh Secti B at Ardingly, Miss M. Reader's Reeves Fairy Lustre, chestnut ma foaled 1961.

7 An excellent example of a successful Arab-Welsh cross: Lemingt May Day, by the highly bred Arabian Stallion, Rukuban out of a lit Welsh mare, Winkie. Lemington May Day was a great winner; includi the supreme Championship at the Ponies of Britain summer show.

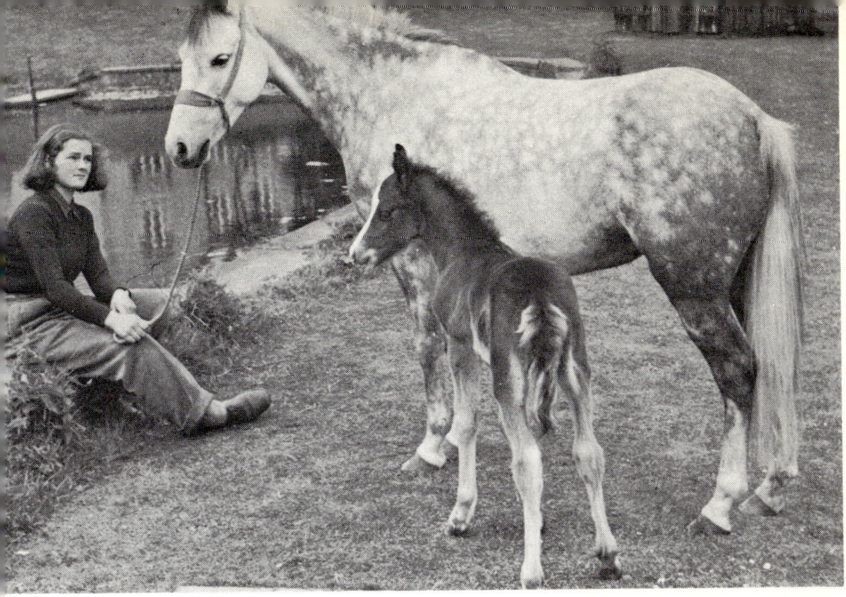

A Friend of the Family. Gillian Cuff with the Welsh pony Downland ilac and her foal, Downland Whitsun, only a few days old.

Treharne Tomboy, owned by Lt-Col. C. Rosser John and bred by rs D. Egerton. This Welsh Mountain pony stallion has superb movement and is a consistent winner.

10 The stallion, Kingsettle Zecchino, many times a champion, driven
Mr A. Robinson for his owner, Mrs Howell, who is with him, at t
National Palomino Championships at the National Pony Show.

11 A popular Spotted Pony stallion, Miss V. M. Willoughby's Fa
Knight of Nashend.

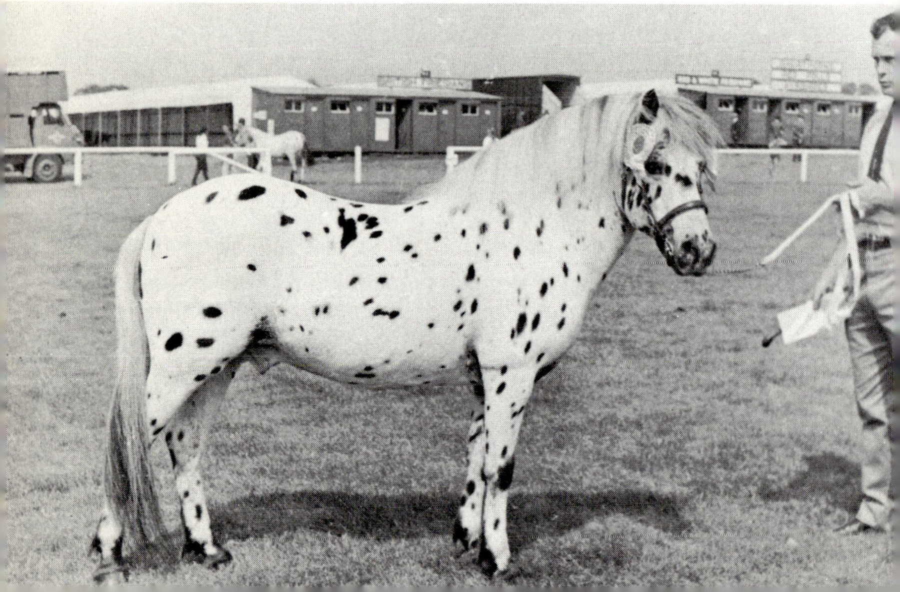

This little Shetland—we do not know his breeding—seems to be joying his day out. He and his nicely decorated caravan look very gay.

Miss S. E. Ferguson showing the black Shetland stallion, Harviestoun ısko, later owned by Miss Nan French. Good-tempered as well as ndsome he won the Shalbourne Cup at the National Pony Show and ıny other awards.

14 Miss S. L. Brook's lovely black New Forest mare, Gosden Nicolet foaled 1961. First at the South of England show at Ardingly 1969, a many other shows.

15 The Highland stallion, Rhum Bloodstone, born on the Isle of Rhu Originally the property of the late Mrs K. Warren, he is now owned Mrs C. M. Smith, who was stud groom at Mrs Warren's Calgary St for many years. This Highland stallion is a first-class ride. He is he ridden by the New Zealand horsewoman, Miss Margaret Jones.

The four-year-old Exmoor pony mare, Crystal, bred by Miss Rose ·cett Green, is by Crackshot II. She is running in a field near the house the edge of the moor with her first foal.

The Fell pony, Dene Beauty Model. This beautiful champion mare is ·ned by Mrs Cave-Browne-Cave, and was bred by Mrs Newall. Note ·e powerful body, matching limbs and fine head.

18 Mr E. Blockley's winning Dales stallion, Stainton Prince at the Poni
of Britain Stallion Show, Ascot.

19 A champion Dartmoor stallion, Hisley Woodcock, owned by Mrs S.
Jones and Miss P. M. Roberts. He is bay in colour and is by Jenson out
Dunnabridge Wigeon.

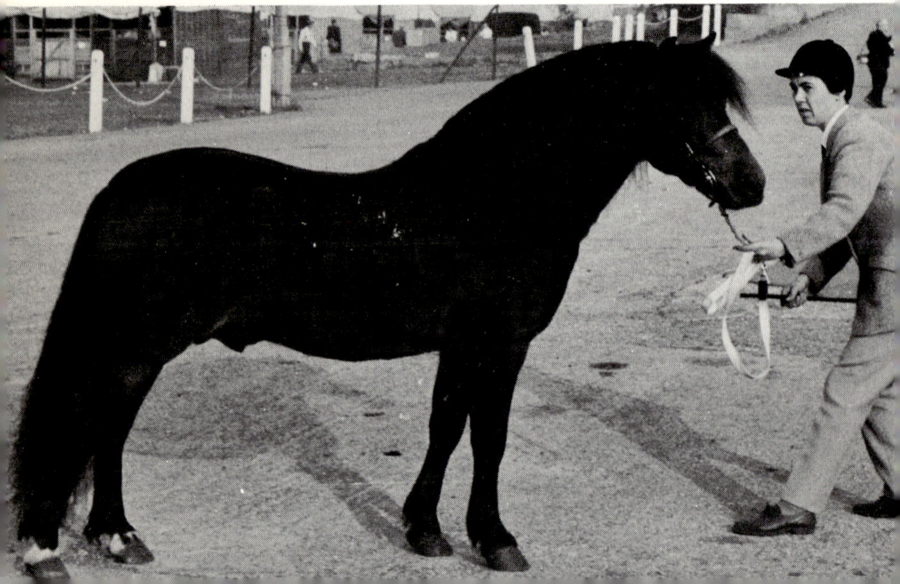

A champion Connemara pony: Mrs G. Foster's Carna Caroline. ɩampion at the Ponies of Britain Show at Peterborough, 1965. A very ɛful type of pony.

The inimitable Pollyanna when owned by Mr A. Deptford and ridden Aly Pattinson. She is by Bwlch Valentino out of the equally famous etty Polly. Note the way she holds her head, the length of rein, the ength, proportions and grace of her body and limbs, which, combined th her excellent temperament, enabled her to become a very good rformer in various events and across country, as well as being a great ampion show pony.

22 Shandon, 13.1½ h.h. a grey gelding by Vagabond, owned and ridd by Miss S. Commerford. He was Children's Pony of the Year at the Ho of the Year Show at Wembley in 1969.

23 Winner of both the novice and open championships at the Briti Show Pony Society's Southern Championships at Ardingly and ma other awards, Dr and Mrs Gilbert Scott's four-year-old chestnut ma Bennochy Ailsa Craig. (First in novice and open not over 14.2 h.h.)

Ireland is Mr Sean Killeen, 4 Nunn's Island, Galway. An important breed show is held each year on the third Thursday in August at Clifden, where many lovely ponies are sold to all parts of the world. Classes are also held at the Royal Dublin Society's Horse Show in August, as well as at many fixtures in England.

They are now being quite extensively bred in both countries, and the English Connemara Pony Society holds its own breed show annually, usually at the end of July. Classes are arranged for every age and sex of registered Connemara and include two ridden classes, open and novice, and a working pony class with jumps. Lieutenant-Colonel R. F. H. Kidd, the President, has presented a challenge cup for the best gelding in the show, and other challenge cups are awarded in every class. There is also an interesting group class for three ponies of any age or sex, all sired by the same stallion. He himself does not appear, only his progeny, but Mrs Cheyne has presented a challenge cup to be awarded to the stallion whose group is adjudged the best. All the ponies entered at this show must be registered and all owned by members of the British Connemara Society.

## Chapter 9

# THE DARTMOOR PONY – INTELLIGENT AND GENEROUS

Another breed which has made a most important contribution to the pony world is the Dartmoor, whose stamina, intelligence and strength is such that he is invaluable as breeding stock and also as a child's all-round pony, capable of anything from dressage to jumping surprisingly large obstacles – in fact, for his size he is a remarkable jumper.

The moorland streams and the valuable minerals which contribute to the strength of moorland stock, the herbage, the invigorating air, and the tors, partially covered with clitters or small rocks and stones, the general up-hill and down-dale going with occasional stretches where the ponies' natural fleetness of foot is given a chance to develop, all combine to produce the qualities and the potentialities inherent in the true Dartmoor. These are, predominantly, strength, sure-footedness, the ability to think for himself – this includes the willpower necessary to deal with the elements on this wild moorland! – hardniess and a very healthy constitution.

The Dartmoor's glorious and protective mane and tail, his fine, small head, broad brow and generous, observant eye, his beautifully formed little prick ears, strong but not small hocks, equally strong back and quarters, good shoulders and very well-made feet are his essential characteristics and should be looked for when you are purchasing a pony. He must not exceed 12·2 h.h. and although bay, black or brown are preferred, there is no colour bar, except in the case of piebalds and skewbalds. Excessive white is not encouraged. The secretary of the Dartmoor Pony Society is Dr E. L. Roberts of Lower Hisley, Lustleigh, Newton Abbot, Devon.

Before describing some of the achievements of the pony today and his use in crossing we will glance back at his varied history and consider the points, the circumstances and indeed the work which has made him what he is. The late Lady Wentworth says in *The Book of the Horse:* 'The pony stock of the British Isles seem to be descended from the small nimble Chariot ponies (seen on pre-Roman coins). They were probably universally distributed in a semi-wild condition, but the ponies of Wales, Cornwall and the Southern Counties would be those which would be most likely to be well-bred, possibly from Arabian root stock.

'English pony stock was undoubtedly "improved" by Arab immediately after the Roman conquest, as the Roman Auxiliary Cavalry brought with them horses of both Spanish and Arabian type. We must not forget that these Dartmoor and the rest are not ancient types but artificially improved ponies based on Arab infusions into Celtic stock. There is therefore no such thing as a pure Exmoor or Dartmoor or pure Welsh pony.'

The late Miss Sylvia Calmady Hamlyn, who was very closely associated with the breed for over 50 years and did an immense amount of invaluable work in connection with it, has many interesting remarks to make in her *Short History of the Dartmoor Pony and Its Society.* She knew many of the moorland folk whose experience dated back to the middle of the nineteenth century and she tells us that 'It seems most probable that the breed originated in the "little British horses" driven with their owner, the "Walli" (Welsh) into the mountains and forests of Dartmoor, Exmoor and Wales and their subsidiary moors by the steadily advancing Saxon conquerors.

'And from that period, the breed running on an entirely unenclosed Moorland – covering 122,000 acres as at present reckoned – must clearly have been subject at every period to alien blood, sometimes intentionally added, sometimes by the turning out or even the escape of stallions on to some quarter of the Moor.

'In Eastern England in the days of the Danish invasions, many horses (stallions) were left behind and would have been used in the locality. The Danes sailed far up the Tamar and other rivers and such a happening is possible in the West. Favourite war horses – Arabs or Barbs – were from time to time brought back by their owners returning from the Crusades.

'All this is, of course, conjecture but it must be remembered that throughout the centuries the aim has been to obtain a better animal, to improve the breed and not merely to keep it static – good or bad.

'In 1905 the National Pony Society set out to improve the various native breeds, many of which had deteriorated in type. Until the first world war various crosses were tried by private owners and by the Ministry of Agriculture, some with indifferent results, some bad and two very good.

'The late Captain Oscar Muntz, one of the men appointed in 1905, brought various pony stallions into the district for experimental purposes. Mikado, a 13 h.h. Fell, sired a useful pony but without the Darmoor quality, and left much hardy stock and some prizewinners, shown as registered Dartmoors. The most outstanding of Captain Muntz's successes, however, was a 12 h.h. Welsh Mountain stallion, Punchinello, and this pony, after populating the Tavistock moor area, went to the Bodmin Moors where he achieved similar success and died there. As late as 1926 he was a tradition on the Cornish moors and sired some beautiful show ponies.'

'In about 1949 the National Pony Society gave the first Betting Board grant to the Dartmoor, and an exhaustive list of Premiums could therefore be offered at the Prince Hall Show and produced a very good entry which in the opinion of judges has improved yearly in quality.'

Miss Calmady Hamlyn herself had an unerring eye and did an immense amount for the breed, running the Society for a long time at her own expense and coping with innumerable difficulties. She would have been immensely satisfied to know that some of the stock she bred with such care has in its turn helped to produce supreme champions of great value, and that the Dartmoor pony is firmly established as one of our leading breeds, often gaining the supreme championships, beating the best of other breeds.

One of today's supreme champions, Hisley Woodcock, owned by Mrs S. E. Jones and Miss P. M. Roberts, is the sixth generation of an unbroken line of champions, tracing back to The Leat, who was foaled in 1918. Hisley Woodcock is by Jenson (by Pipit out of the incomparable little mare, Jenny VII, well known to the author) and his dam was Dunnabridge Wigeon. Pipit is by Jude out of Peewit III, who at the time of writing is 31 years of age and has retired from breeding. But, to quote her owner, Miss Jean McMillan, to whom she was given by the breeder, Miss Calmady Hamlyn, she is very contented as she looks after her grandchildren each year and finds that just as satisfying. Jenny was by Jude out of Dora, another great pony of the past. Jude's dam was Juliet IV by The Leat.

Hisley Woodcock, 11·2 h.h. and of immense strength and presence, has an amazing record when competing with other mountain and moorland stallions for the two principal mountain and moorland championships of the season. He won the Shalbourne Cup at the National Pony Show in 1969 and the Merlin Cup at the Ponies of Britain Stallion show at Ascot in 1967 and 1968. Like his sire, Jenson, he has won the supreme championship at the Dartmoor Pony Show three times.

There are, of course, a number of other beautiful stallions and mares of considerable value to the breed and to their owners on Dartmoor, in Somerset, Berkshire, Kent and many other parts of the country as well as in Scotland, but I have quoted this one in particular as his success combined with his ancestry is of such interest.

The lovely mares, themselves great show ring champions, who are now producing uncommonly attractive foals, make an immense contribution to the history of the Dartmoor pony. Many have gone or are going, overseas, as are the stallions, and there is little doubt that their individual stories would be of real interest, as the Dartmoor is such a highly individual pony.

One old mare, Mountain Sweet, has an astonishing record in that every one of her foals has been a prizewinner at major shows, although she herself is no beauty. In addition to breeding pure-bred foals she has been used with pony stallions such as the Thoroughbreds Ardencaple and Gay Presto, the Arabs Blue Domino and Blue Halo, and the pony-bred sire, Bwylch Zephyr. These first-cross foals have grown into wonderful ponies, winning classes and championships in hand and under saddle at an impressive variety of leading shows against strong competition.

A part-bred Dartmoor, Lady Willingdon's Kilbees Blue Jenkin, by Gay Presto (T.B.) out of the registered Dartmoor mare, Vean Blue Jennykin, has also sired some noticeably good progeny. Others, such as Vean Zenith by Bwlch Zephyr, out of Mrs Robinson's exquisite little Dartmoor, Vean Zelinda, have added glory to the breed and, indeed, the list is so long that it is impossible to enumerate even a tenth of it. The fact that the Dartmoor has so often contributed so greatly to the production of the very best type of children's riding pony is indisputable. One wishes that it was possible to mention individually the many excellent studs of these ponies, both on their native moor and up-country.

The value of Dartmoor blood in the production of high-class ponies of various sizes has caused breeders to form 'The Part-bred Dartmoor Register', of which the Secretary is Commander J. Palmer, River Hall Farm, Biddenden, Kent. Its object is to assist and encourage owners to breed a quality riding pony not exceeding 14·2 h.h. at maturity, with real pony character, good movement and temperament.

To qualify for entry a fully registered pony may be out-crossed to a Thoroughbred, Anglo-Arab, Arab, or a riding pony stallion registered in the National Pony Society Stud Book or appearing in the riding pony stallion register. If a Dartmoor stallion is used the mare must be entered in one of the above, or else in the approved mare register of the National Pony Society. Not less than 25 per cent Dartmoor blood will qualify for entry, and not more than 25 per cent other native blood may be used.

In all cases Dartmoor registration numbers must be on the entry form or the entry will not be accepted. This register, or society, is run by a small, experienced committee and presents a rosette at a number of the leading shows.

The Dartmoor blood in many of our leading children's ponies is worth noting. For example, the Royal Welsh champion and winner of the class for children's ponies not over 14·2 h.h. at the Royal International Horse Show in 1969, was quarter Dartmoor. This was Downland Ballerina.

## Chapter 10

# THE TOUGH DALES PONY

It seemed for a while as if the hardy, extremely strong Dales pony would cease to exist, as he was more of a working pony, suitable to carry loads and to be driven, than a riding pony, and not much attention was given to breeding him. Now, however, the situation has altered, he is being more widely bred, has his own society and takes part in various horse shows, in addition to the breed show held in conjunction with the Durham County Show.

The Dales pony's notable abilities came to a great extent from the country in which he was bred, i.e. the North Riding of Yorkshire and parts of Northumberland and Durham. His ancestors worked for a long time in the Dales, east of the Pennine Range of mountains, which he climbed up and down on narrow, natural, often difficult, tracks, transporting lead from the mines in Northumberland and Durham to the docks. He carried it on either side of his body, the total burden probably amounting to as much as 16 stone, and covered 240 miles in a week. Whether the Dales pony which is bred today could do this work is a questionable point, but it is not necessary that his resources should be exploited, as his purpose in life is quite different, and indeed he is used for many other things.

One of these purposes is breeding. As the Dales Pony

Society will tell you, the Dales mare, if crossed with a Thoroughbred stallion, will produce excellent weight-carrying cobs and hunters with substance.

The ponies are admirably suited to the hilly, Northern Dales farms, where their sure-footedness and strength enable them to do all the farm work and where a bigger horse would be at a disadvantage. Their tremendous stamina makes it possible for them to carry the shepherd who has large sheep areas on the fells and moors. They can stand great cold, and, owing to generations of hereditary soundness, they are seldom sick or unable to work.

They make excellent trekking ponies, and true 'family' ponies, able to carry adults, thanks to their short backs and strong loins, or children. Their temperament is placid and reliable and they are particularly good for nervous, perhaps heavy children at Pony Club rallies or out hunting, as they are clever and willing jumpers but do not 'hot up'. Another considerable advantage is their excellent behaviour in traffic and they are indeed good harness ponies, either for the small trader, or the member of a driving club.

The true Dales pony should have a neat head and small, well-placed ears and he should be free from coarseness about the jaw and throat. His back should be short, with strong, powerful loins and his quarters and second thighs well developed. The feet are, as a result of sound breeding, hard and well shaped, and the legs and joints are excellent, with plenty of flat bone. There is a certain amount of hair on the heels and this should be fine and silky.

It is vital that the action of the Dales should be straight and true, and to quote the description supplied by the Society itself: 'They are good movers, really using their knees and hocks, which gives the onlooker, when these ponies are in action, the impression of sound, rhythmic movement and power.' In height they should not exceed 14·2 h.h.

Mr George H. Hodgson is secretary of the Dales Pony Society, and his address is Ivy House Farm, Hilton, Yarm-on-Tees, Yorkshire.

## Chapter 11

# THE STURDY EXMOOR PONY

The Exmoor pony, extremely sturdy, intelligent, and with the charm of character, individuality and determination which might be expected from his long sojourn in the beautiful demesne of Exmoor, is often the subject of debate. Is he indeed the descendant of the native British wild horse or, as some people think, is the modern idea that this pony is a pure, original breed entirely without foundation? It has been suggested that Norwegian blood was introduced at some time.

Let us study various opinions. Mr R. S. Summerhays states in *The Observer's Book of Horses and Ponies* that the Exmoor pony is a descendant of the native British wild horse and is believed to be an indigenous animal that has been preserved in its aboriginal state from the earliest times to the present day.

In *The Exmoor Pony*, published by the Exmoor Pony Society, we learn that Exmoor 'is a solitary place and the wild animals which live on it need to be hardy in order to survive the winters. In days gone by, before man exterminated the wolf, the grass eaters had to defend themselves against their enemies as well as the climate. Nature saw to it that these grass-feeding animals, such as the wild Red Deer and the Exmoor pony, were so clothed as not to be readily seen and the coats of both are beautifully camouflaged to harmonize with the background of the moorland on which they live.

'During the Ice Age, Exmoor was one of those parts of the British Isles which was left uncovered by the great glacier, with the result our ponies survived. Many of us in the Exmoor Pony Society, aided very much by the research work carried out by Professor and Mrs Speed at the Royal (Dick) Veterinary College in Edinburgh, believe as a result that the Exmoor pony is unique amongst native breeds, and has been preserved in its original state from the very earliest times right down to the present day.

'The Exmoor pony is, then, one of the hardiest ponies in the world, and this is one of the reasons we believe them to be so important to horse breeders. They provide the finest foundation stock to be found and so, for this reason, a touch of Exmoor pony blood in any hunter or working horse ensures sense. You have all heard of horse sense, and pony sense is the same thing, only more so. The Exmoor pony has great powers of endurance and it is agile and so can avoid all kinds of troubles in awkward situations. When crossed with a Thoroughbred the foals inherit the best characteristics of both sides and, in fact, one Grand National winner sprang from Exmoor pony stock.

'They can carry extraordinary weights which would be beyond the capacity of many others, and it is no uncommon sight to see a large farmer shepherding on the hill, knowing full well he will never tire his pony, which will provide him with a safe and worry-free ride while he is able to turn his full attention to his sheep and cattle.'

Whatever may be the correct answer to the original breeding of the Exmoor pony, there is no doubt that wise crossing, using a good registered Exmoor sire or dam, is very rewarding. In several instances the use of the right type of Arab stallion on an equally nice Exmoor mare has produced most excellent riding ponies, not so much Arabs but greatly improved Exmoors; and it is certain that if one goes back for quite a long time one will find evidence that Exmoors were crossed with Arabs.

A letter from T. F. Dale, published in *The Livestock Journal* of January 1920 is headed 'On T.B. and Arab Crosses with Native Ponies', and states: 'I remember when I first hunted on Exmoor, after returning from the East, being struck with the Arab character of some of the Exmoor ponies as they stood and gazed on the hills as the chase swept by. . . . I recollect the present Earl Fortescue showing me one which he had bred, using, I think, the Duke of Connaught's Arab stallion which then stood at Castle Hill and one of his own Exmoor mares.

'This pony was used for tufting.* He was a bay pony of great quality and sufficient substance to carry Lord Fortescue through a morning's tufting. I went with him on purpose to watch the pony and a better pony for tufting there could not be – handy, quick and full of courage. After about two hours' hard work he was not in the least distressed.'

To bring the record up to date, when I had the pleasure of judging a large number of classes of all kinds of ponies at the Canadian Pony Breeders Two-Day Show in Ontario, not only was I impressed by the quality of the pure-bred Exmoors there (I did not judge them, just observed from the ringside), but also by the performance, appearance and popularity of those ponies resulting from an Exmoor small Thoroughbred cross. A pure-bred registered Exmoor, Mrs B. McInroy's

* In tufting experienced hounds are taken to a covert to rouse the stag and get him to leave the covert and make for the open moor. The tufters are then held up and the rest of the pack, which is temporarily kennelled, is laid on the line. There is often a delay of 30 minutes, which gives the stag a chance to make his getaway.

Early Bird, ridden by Lucy Alderson, won the small green hunter championship, as well as other classes at this show and also put up a most creditable performance at the Royal Winter Fair in Toronto later the same year, jumping in first-class style against very strong competition. This speaks well for the potentialities of the breed, which has its own society in Canada – a branch of the Exmoor Pony Society in England. As the ponies grow bigger in Canada than they do on Exmoor the height limit is 13·2 h.h., as opposed to 12·2 h.h. for the mares, and 12·3 h.h. for the stallions at maturity, which is the case in England.

Lieutenant-Colonel V. C. A. Munckton of Capland Orchard, Hatch Beauchamp, Taunton, Somerset is secretary of the British society, which offers its members many advantages. It keeps a very useful register of first-cross Exmoors, the rule for inclusion being that one parent must be either Thoroughbred (General Stud Book) or a registered Arab, or else a registered pony from one of the other mountain and moorland breeds. The other parent must be a registered Exmoor. The cross with a good Thoroughbred pony stallion has brought some quite spectacular successes.

The pure-bred Exmoor always has a mealy muzzle – the colour of oatmeal. His forehead is wide, with prominent, unusual and beautiful eyes, often referred to as 'toad eyes' because the lids are slightly hooded. The ears are short, thick and pointed, and they, too, have a mealy colour inside. The nostrils should be wide and of generous proportions.

These ponies may be bay, dun, or brown, but under no circumstances are white markings, or white hairs anywhere, accepted. The coat is unusual and extremely useful as a protection against the weather. It is harsh and springy, carrying no bloom in the winter, but in the summer it is close and hard and shines like brass. Foals have a thick, woolly undercoat, with a top coat of long, harsh hairs which acts as an overcoat to keep the rain from penetrating.

The sire and dam of ponies whose owners wish to register them must be entered in the Stud Book; they must be sound and have been inspected and passed for registration by two of the Society's panel of inspectors, also branded with the Society's star and herd number on the near shoulder, and the pony's own number in the herd on the near pin.

## Chapter 12

# THE FELL PONY; STRONG BUT GRACEFUL

When one considers his powerful make and shape one must acknowledge the grace of the Fell pony. His natural habitation is the Western area of the Pennine Range of hills; wild country, where only the hardiest animals can stand the climate. Like his brother, the Dales pony, the Fell pony was used many years ago – in fact, during the latter half of the last century and possibly the beginning of this one – to carry the lead from the mines to the docks on Tyneside. Droves of twenty in charge of a man on horseback were kept at a steady walk, bearing 8 stone of lead on either side. By way of contrast they also competed in trotting races arranged by the farmers of their own particular district, and these races were also featured at some of the shepherds' meets. In those days they were bred to trot at great speed.

Now the Fell has proved himself as a good riding pony, and he may also be seen going very well in harness at the rallies of the British Driving Society. These intelligent ponies adapt themselves to their various jobs extremely well – hunting, shepherding, long distance rides, trekking and combined training which, of course, includes elementary dressage.

Probably the most famous horse of Fell ancestry today is Miss Anneli Drummond-Hay's Merely-a-Monarch, a consistent winner equally brilliant in horse trials (he won the

Burghley Three-Day Event in 1961, giving a particularly remarkable performance in all three phases – dressage, speed and endurance, and show-jumping – as well as the Badminton Horse Trials in 1962) and in international show-jumping. He has retired from horse trials now and devoted his talents to jumping, and at the time of writing is still winning all over the world. His great grand-dam was a Fell.

If you wish to cross a Fell with anything else a Thoroughbred is usually recommended as the most consistently successful. A Fell stallion and a hunter mare have also produced good stock, in the shape of a very useful and hardy type of small hunter. The offspring of the Fell and the Thoroughbred almost invariably jumps well, in the show ring or across country.

Miss Peggy Crossland, secretary of the Fell Pony Society at Packway, Windermere, Westmorland, tells me that she 'owned for 27 years a wonderful gelding, three-quarter Thoroughbred, quarter Fell, who seemed to combine the good points of both breeds. He was 14·2 h.h., a good-looking, bright chestnut; very handy, sometimes lazy, sometimes a real handful, but in a tight corner the sensible Fell temperament usually came out on top. He was also a very faithful friend and a really lovable character.'

Miss Crossland also writes, 'I had a letter today from a man who owns a Thoroughbred × Fell stallion. He says "I would like to tell you of the success I have had with The Black Prince. Last season he sired some really beautiful foals and he was never placed lower than third in the show ring. At one point he almost became a film star. Alas, he was 1½ hands too small for the man who was to have ridden him in the film, nevertheless to us he is a star and a most memorable one." This stallion is registered in the Riding Pony Section of the National Pony Stud Book.'

The height limit for a pure Fell is 14 h.h. and they usually range from 13 to 14 h.h., which is not quite as tall as the Dales (14·2 h.h.). At one time there was no distinction between the two breeds, but that was a long while ago and is no longer the case. They have been bred differently during more recent years, the Fell pony remaining remarkably true to type. The Fell Pony Society and the breeders take particular pains to ensure that no alien blood is introduced by accident or design, as they are most anxious to ensure that the breed is pure and does not lose the characteristics for which it is noted.

The following description of the Fell pony and the scale of points issued by the Society may interest you.

COLOUR; black, brown, bay and grey preferably, with no white markings, although a star, or a little white on the foot is allowed.

HEAD; small, well-chiselled in outline, well set-on, forehead broad, tapering to nose.

NOSTRILS; large and expanding.

EYES; prominent, bright, mild and intelligent.

EARS; neatly set, well-formed and small.

THROAT AND JAWS; fine, showing no signs of throatiness or coarseness.

NECK; of proportionate length, giving good length of rein, strong and not too heavy, moderate crest in case of stallion.

SHOULDERS; most important, well laid back and sloping, not too fine at withers, nor loaded at the points – a good, long shoulder blade, muscles well developed.

CARCASE; Good strong back of good outline, muscular loins, deep carcase, thick through heart, round ribbed from shoulders to flank, short and well-coupled, hindquarters square and strong with tail well set on.

FEET, LEGS AND JOINTS; feet of good size, round and well formed, open at heels with the characteristic blue horn, fair sloping pasterns not too long, fore-legs should be straight, well-placed, not tied at elbows, big well-formed knees, short cannon bone, plenty of good flat bone below knee, eight inches at least, great muscularity of arm.

HIND LEGS; good thighs and second thighs, very muscular, hocks well let down and clean cut, plenty of bone below joint, hocks should not be sickle or cow-hocked.

MANE, TAIL AND FEATHERS; plenty of fine hair at heels (coarse hair objectionable), all the fine hair except that at point of heel may be cast in summer. Mane and tail are left to grow long.

ACTION; walk, smart and true. Trot well balanced all round, with good knee and hock action, going well from the shoulder and flexing the hocks, not going too wide nor near behind. Should show great pace and endurance, bringing the hind legs well under the body when going.

GENERAL CHARACTER; The Fell pony should be constitutionally as hard as iron and show good pony characteristics with the unmistakable appearance of hardiness peculiar to mountain ponies, and, at the same time, have a lively and alert appearance and great bone.

SCALE OF POINTS; Height and colour, 5; head, nostrils, eyes, ears, throat and jaws, neck, 10; shoulders, 15; carcase, 20; feet, legs and joints and hind legs, 25; action, 25; general characteristics, 100.

There is considerable demand for the registered Fell pony, particularly for those which have been broken to ride or drive. I was delighted to see some very fine representatives of the breed in Canada.

## Chapter 13

# THE BEAUTY OF THE HIGHLANDS

The colours of this immensely powerful breed are in fact as beautiful as they sound – blue dun, silver dun, mouse dun and many others. The Highland pony is also to be found in lovely shades of grey and cream and sometimes a chestnut with a silver mane and tail appears among those from the Western Isles.

To quote from the booklet, *The Highland Pony*, by the late Mrs K. Warren, herself a great breeder of champions. 'The eel or dorsal stripe, is a usual mark of distinction, as are the zebra markings on the legs which are seen on most dun-coloured ponies and occasionally on greys, too. The yellow dun colour, said to be the original, is perhaps the most interesting. One rarely sees it in its true colour which has a great deal of black in it. If one separates the hairs down to the skin they will look smoky black at the roots, changing to golden towards the tips. While the coat is changing, especially in a young animal, it loses its brilliant golden colour and becomes dusky looking, and often one can see zebra stripes in it, particularly on the neck and withers. These disappear as full yellow coat comes in.

'The colour of the full coat should be vivid gold, not a light bay or chestnut colour, and the points should be dark seal or black. "Points" in this case being the tips and edges of the ears, the face, mane and tail, dorsal stripe, zebra markings on the forearms and hocks, and legs up to or over the knees and hocks. The horn of the feet should be dark and the skin of the body pigmented. There should be *no* white marks. Some yellow duns have almost mealy noses, and as they advance in years the dark face and mealy nose becomes more pronounced. Many of these ponies have the dusky shading and stripe over the wither like a donkey. If any cross-breeding is introduced the true yellow dun colour is lost.'

The Highland pony comes from the Western Isles and Outer Hebrides and the Mainland of Scotland. It is claimed that he is a descendant of the second oldest race of horses that can be traced to have existed as far back as 60,000 B.C. From remains found in fossils it seems that the ancestor of the Highland pony, then about 11 h.h., flourished in Britain before the Ice Age.

There are two distinct types of Highland pony, the smaller pure-bred is from 13 to 14 h.h., and the larger, the Highland Garron, is 14 to 14·2 h.h. The smaller ponies have velvety noses with wide nostrils, generous eyes often golden-hazel in colour, quality heads wide between the eyes, and very small, prick ears. Their legs are strong and of a length proportionate to their height, with closely placed hocks, the feet open and round and composed of excellent horn and the pasterns sloping but not too long. There should be no hair on the legs except for small, silken tufts on the fetlocks. The hair of the mane and tail must be pleasant to the touch, not harsh or wiry.

The larger ponies have tremendous strength in relation to their size but sometimes you may find evidence of cross-breeding in their disproportionately larger heads, hairy noses which are often pink, and far too much hair on the heels and over the coronet. Some of these ponies originate from the crossing of the pure pony with cart horses to get an animal suitable to work on the larger farms and deer forests. Selective breeding from the progeny of these ponies has been carried out.

The smaller ponies are also often very sturdy and are popular in the deer forests as it is easier to load a stag on to one of them than on to a larger animal. They are very good at the more practical jobs connected with a farm and make splendid riding ponies, suitable for most branches of equitation, hunting, jumping, hacking and exploring little-known wild country. Sure-footed, sensible and, indeed, clever, they can be relied on to make their way safely along mountain paths or boggy, trappy going.

Their temperament makes them most interesting, as they share their gift of memory with the elephant 'who never forgets' – therefore correct training and kind, firm treatment reap rich rewards, whereas the reverse certainly does not pay off! Once they become attached to anyone – and they do not do this immediately – they have great trust in them and they are sensitive and intelligent.

I will again quote Mrs Warren's booklet *The Highland Pony,* in which she described the mares who, as she says, are very competent mothers. 'One old Western Island mare, belonging to a friend of mine in Mull, used to hide her foals for the first week after birth, as do the deer. I owned her granddaughter for several years, and she was very jealous of her foals and even hated us to look at them. She used to take her foal to the further corner of the field and then stand right across it, never grazing while it slept but watching all the time.

'It is one of the most marvellous and impressive sights if one is ever lucky enough to be present when a Highland mare foals, the perfect work of nature and the dignity and grace and super sense of the young mother is indeed a wonder of wonders to watch, and quite beautiful – the mare looks her loveliest, her eyes are startlingly bright and her head looks more shapely and alert in expression than usual. It would be difficult for the most hardened Atheist to deny the existence of a Divine Creator, were he witnessing such a scene. Once the opportunity was afforded to me to give my doctor husband, who has brought many human babies into the world, the chance to see a little Highland pony mare have her first foal. I could see he was absolutely lost in wonder and admiration at the whole proceeding, but said nothing till we sat down to breakfast together when all was over and the new baby guzzling her first meal, then he said "and *we* think we are clever!"

'There is no more charming sight on a sunny May morning than a bunch of Highland mares with their prettily-coloured foals lying in fresh green grass. One old yellow dun mare of mine put her little cream daughter to sleep in view of our morning room window, and as it slept I saw a little grey squirrel sitting beside it, with its little paws folded across its chest and its little beady, bright eyes watching the foal with intense curiosity. . . . It is always best to leave the foal with its mother until nearly a year old, as like Highland cattle they do better than if weaned earlier like other breeds. Many Highland mares will only breed every other year.'

In the past there was quite a large number of famous studs of Highland ponies, almost all the large estates having their stud of ponies, but, alas, not many remain today. The Rosehaugh Stud bred Jock, owned and ridden by the late King George V – he carried him up to the day of his last illness and was immortalized in a painting by the late Sir Alfred Munnings. Jock was sired by Glen Bruar, whose stock achieved much fame.

Not many of us realize that Troop 1A Squadron and

another Troop formed in the Isle of Tiree were both originally mounted on Highland ponies and formed a part of the Regiment called the Scottish Horse, raised by the Duke of Atholl.

Correctly schooled and ridden, the Highland pony makes a most excellent riding pony and his strength and general usefulness is worth remembering. He is also a good doer as well as a good friend.

Mr J. McIldowie is secretary of the Highland Pony Society at 51 High Street, Dunblane, Perthshire.

# Chapter 14

# LUNDY ISLAND PONIES

In a book dealing with our native breeds and their crosses it is only fair to refer to two types of pony which, although they may be said to run in herds rather than breeds and as such cannot be eligible for registration in the National Pony Stud Books as their actual breeding is somewhat mixed, still have their own acclaim and value, and are, indeed, an example of crossing. These are the Lundy Island pony and the Quantock pony (see page 93).

The first of these, the Lundy Island pony lives on this little Island in the Bristol Channel, and at the time of writing are probably about twenty in number. Some New Forest mares and fillies were taken there in 1928 by the naturalist, Martin Coles Harman, together with two stallions.

One of these was a Thoroughbred and his foals could not stand up to the severe winter conditions but a small Welsh cob stallion was used more successfully. Latterly fresh outside blood was needed and a young Connemara stallion, Rosenaharley Peadar 4561, was purchased and has sired good, strong foals.

The predominant colour of the Lundy Island ponies is of varying shades of dun, although roan is not uncommon, and this stallion is also dun – mouse-coloured in winter and golden in summer.

Each year in June the yearlings not kept for breeding purposes are shipped on a fishing ketch to the mainland and sold at Barnstaple market. These wild ponies react very well to kind and patient handling and some, such as the Olympic Three-Day Event rider, Mr Bertie Hill's first jumping pony, Lundy, have proved their worth, their usefulness and ability, in no uncertain fashion.

## Chapter 15

# OUR VERSATILE FRIEND, 'THE FORESTER'

The varied and interesting history of the New Forest pony has been carefully traced by Sir Berkeley Pigott in his excellent little booklet published by the New Forest Pony Breeding and Cattle Society, of which the secretary is Miss D. Macnair, Beacon Corner, Burley, Ringwood, Hampshire.

Sir Berkeley points out that King Canute's law of the Royal Forest, proclaimed at Winchester in 1016, mentions large numbers of cattle and beasts of the Forest, such as horses, buffaloes, cows and the like, living in the Forest. But it was not until 1877 that the Verderers, who had been appointed to protect the rights and privileges of the Commoners, were given power to make bye-laws on various matters, including the complete control of all stallions turned out in the Forest.

He goes on to tell us that 'On the 10th February, 1891, the Association for the Improvement of the Breed of New Forest Ponies was inaugurated at Lyndhurst with the object of awarding each year prizes and premiums to stallions after they have been passed by the Verderers to run in the Forest. Prior to this there is little record of any organized attempt to improve the breed, with the exception of a grey Arab stallion, Zorah, which stood at New Park from 1852 to 1860, having been lent to the Verderers by the Queen [Victoria]. There can be little doubt that there had been previously a great many crosses with Arabs, as to this day a very noticeable Eastern type reappears.

'The late Lord Arthor Cecil, believing that the best method of introducing fresh blood into our mountain and moorland ponies was by the introduction of stallions of kindred breeds, commenced importing in 1893 a number of stallions from the Highlands, from the Island of Rhum, from the Fells of Cumberland, from Dartmoor and Exmoor and later, some choice strains from the mountains of Wales. At the stallion show in 1914 out of the 121 ponies exhibited nine were Welsh, five Dartmoor, and four each from the Highlands and Exmoor. Yet owing to the mysterious power of nature to grind down and assimilate all these types to the one most suited to the land, New Forest ponies are still fairly of one type and they are rapidly becoming more so. For the last 30–40 years no outcrosses have taken place.

'The Burley and District New Forest Pony and Cattle Society was formed in 1906. Its chief activity was the holding of an annual show on August Bank Holiday for mares and young stock and the programme also included riding and harness classes, as well as races for New Forest ponies. In 1910 the above Society commenced to publish a Stud Book; a New Forest section has since been incorporated in the National Pony Society Stud Book. In 1959 the New Forest Breeding and Cattle Society decided to publish its own Stud Book; Volume 1 being available in 1960.

'A New Forest pony is one already registered in the New Forest Stud Book or the New Forest Section of the National Pony Society's Stud Book or the Local Record. A pony is eligible for registrations if its sire is a registered pony holding a Ministry licence or a Veterinary Certificate of freedom from hereditary unsoundness and has either been passed by the

Verderers to run on the Forest or for whom a return has been received by the hon. secretary.

'The dam must be registered as above and if bought in-foal a covering certificate from the former owner is required. A pony is eligible for entry in the Appendix to the Stud Book if its sire is registered as above and its dam is a pony running on the Forest known to the Agister and the Council as of pure Forest stock. A colt so bred is only eligible as a gelding. In all cases the pony must be neither piebald nor skewbald and not exceed 14·2 h.h. in height.

'In 1938 the Association for the Improvement of the Breed of New Forest ponies and the Burley and District Pony and Cattle Society were amalgamated under the title of "The New Forest Pony Breeding and Cattle Society". In addition to the Stallion and August shows the Society organizes shows each Spring for non hand-fed mares, i.e. the mares that run out on the Forest throughout the summer and winter; each exhibit is judged in the district which they haunt. Point-to-Point races are held on Boxing Day; the entries having to ride three and a half miles across the Forest from one point to another and being unaware of the course till they reach the assembly point.'

To complete these extracts from Sir Berkeley's book I will add that races, not point-to-points, for New Forest ponies are also held in conjunction with the show in August and auction sales take place several times each year at the Society's own sale yard near Beaulieu Road station. All information is available from the society's Secretary, Miss D. Macnair.

It may also help if I explain that a 'commoner' is one who holds the right of common pasture, and an 'agister' or 'marksman' is employed by the Verderers to 'collect the marking fees of all commonable animals turned out on the Forest. These fees are £1·25 per head for cattle and ponies. There are at present four agisters and each has a distinctive method of marking the pony's tail, supposedly done at the time when the fees are paid.'

The official description of the New Forest pony is as follows:

HEIGHT; not exceeding 14·2 h.h.

COLOUR; any colour except piebald or skewbald.

DESCRIPTION; a good riding type of pony. Head well set on. Neck a little short from throat to chest, but good laid back shoulder giving plenty of length of rein. Short back, strong loins and quarters. Tail well set on though not exaggeratedly high. Good forearm and second thigh, short cannon bones, good feet. Plenty of bone. Straight but not exaggerated action.

TYPE A; Ponies up to 13·2 h.h. Lighter in bone than the larger ones. Suitable as ideal children's hunting ponies, with quality.

TYPE B; Ponies 13 h.h. to 14·2 h.h. with plenty of bone and substance, a strong type of pony able to carry adults. The New Forest pony is the ideal family pony. Though classed as one of the larger breeds it actually fills the gap between the small and the large, as it varies in height from 12 to 14·2 h.h. The average height of these ponies is 13·1 h.h.

Yes, one must agree that the New Forest pony has much to be proud of. He takes part in his own races and long-distance rides, polo and also colt-hunting in the spring, and the older members of this most varied and versatile family compete in veteran classes for ponies 20 years old and over – one of these was won by Mr F. C. Bennett's wonderful old stallion, Goodenough, whose group also won the progeny class on the same day.

At the other end of the scale is the more common and cheaper 'Forester', who will be a good and useful friend to any child. One four-year-old gelding, known to have a charming temperament, was exported to Sweden to a lonely, nervous child, without much money to spend on the purchase of a pony. He literally transformed her life and she in her turn made him into a good ride, attaining quite a high standard of dressage with him. Many New Forest ponies have been exported and a large number are used by members of Pony Clubs in Holland.

This breed has produced more than a few useful part-bred or first crosses, the majority of whom are excellent small hunters ranging from 14 to 15 h.h. Perhaps the most famous part-bred is Royal Show, referred to in Chapter 5. This child's pony was a consistent winner of supreme championships. His grandsire was a Thoroughbred and his granddam a New Forest pony; the filly born as a result of this cross was put to an Arab and she foaled Royal Show.

Another good example is Beacon Peto, by the Thoroughbred, Eastward Bound, out of the New Forest pony, Bettesthorne Kate. He won useful championships as a yearling, grew into an exceptionally fast, bold pony, doing very well in one-day events and as a J.A. jumping pony – he was 14·1 h.h. There have, in fact, been many first-class jumping ponies who were the result of a first cross. Some people prefer to use a Thoroughbred to cross with a New Forest pony, although the Arab has produced some excellent stock, although with the latter it is best to choose a really deep, small-headed mare.

In Denmark New Forest ponies have been crossed with Trakheners and have bred good jumping ponies for teenagers.

Finally the 'Forester' is a grand pony in harness and both singly and in pairs has won or been placed when up against very strong competition in private driving classes.

## Chapter 16

# THE PALOMINO PONY: OFTEN THE PRODUCT OF NATIVE STOCK

It may not be technically correct to include the lovely, almost legendary, Palomino pony among our native breeds but, like the Spotted Pony, it is bred in Britain to a considerable extent and as so many Welsh ponies and quite a few Connemaras are also registered Palominos it seems unfair to leave them out.

Although they are a colour, not a breed, these ponies go far back in history, and they have added their quota to the glamour of tradition, besides giving much pleasure to the artist and the individual who appreciates pageantry. In fact, they probably added greatly to the panache and presence of the famous people who rode or drove them and are immortalized in various pictures.

The Palomino horses, so greatly treasured in America, Spain, Australia and other countries, are not the subject of these comments but rather the ponies of all sizes so often to be seen with their breeders, riders or drivers in the show ring and also on the country roads.

The correct colour of a Palomino is likened to a newly minted sovereign, a shade of gold which, allied to the flowing snow-white mane and tail, presents a breath-taking spectacle. However, this is perfection, which cannot always be achieved, and three shades lighter or three shades darker are acceptable, but not what one might describe as near-chestnut or off-cream.

White markings are not allowed on the body, except where they are the results of injury, only on the face and legs. Manes and tails must be white with no more than 15 per cent of chestnut or dark hairs. Eyes must be dark brown, hazel or black iris, and both of the same colour. Ponies with wall eyes are not accepted for registration.

Because the coat changes in colour, often becoming darker as the young stock grows to maturity, the registers are kept by the British Palomino Society, of which the secretary is Mrs Peter Howell, Cholderton, Salisbury, Wilts. One of these is for animals up to six years old only, and the other is a permanent register for animals over six years. The rules for inspection are as follows:

Animals for which application for registration in the Young Stock Register has been received will be inspected at any time of the year on or after 1st April in their yearling year; up to the 31st August in their five-year-old year by one official inspector, appointed by the Society.

Unless application for registration of young stock has been received by the secretary not less than 28 days prior to the closing date for entries for any show, the Society is unable to undertake that registration formalities will be completed by the date of the show.

Animals for which application for registration is received after 3 August in their five-year-old year will be considered eligible for registration in the Permanent Register only.

Inspection of animals for acceptance into the Permanent Register will be made by two official inspectors, with referee available, during the months of June, July and August only at centres to be selected annually. Owners of registered young stock will be informed early in each year of the selected centres and provided with Permanent Register application forms for the animals believed to be in their ownership which are six years old that year.

The list of centres is also published each year in the Spring issue of *Palomino*, which is the excellent illustrated quarterly magazine distributed free to members of the British Palomino Society.

When cross-breeding with a view to producing Palominos do remember that in addition to trying to achieve the right colour you must also breed for conformation, movement and temperament. The reason for this is twofold (although surprisingly it is sometimes overlooked): (*a*) The good pony will always be useful for a number of different jobs, and will fetch a fair price even if he is not a true Palomino in colour: (*b*) It is important to breed high-quality Palominos, capable of winning in any company, not only in competition with his fellow Palominos. Thus, and thus only will the Palomino flourish and his value and popularity increase.

With a little wise forethought and the choice of a good sire and dam this is not difficult to carry out, and it pays handsome dividends. There are Palomino Shetlands, Palomino Welsh (pure and cross), Palomino part-Thoroughbred, Palomino Pleasure Horses and so on, up and down the line in size.

As examples of breeding which produce good Palomino ponies (remember that we are discussing ponies, not horses – there are many fine stallions from which to breed the larger animal) take Mrs Loriston-Clarke's Bubbly, 13·1 h.h., whose colour has remained so consistently golden throughout his life. He is by a fine, Thoroughbred stallion, Potato, who so many good ponies out of the nice pony mare, Sunday. Or the Welsh Mountain registered Palomino stallion, Rowfant Santiago, 11·2 h.h., by Twyford Grenadier out of Coed Coch

Sianw. Or the supreme champion, Kingsettle Zecchino, 14·2 h.h., by the Palomino stallion, Golden Trail, imported from Wyoming, U.S.A., out of the part-bred Arab, Rosecyra, who was by Craven Cyrus out of Rozanda. Or Drayton Bubbling Gold, by 14 h.h., by Bubbly out of the well-bred children's pony Golden Oriole. Or Wheelers Story Teller, 13·2 h.h., by Turton's Story (G.S.B.) out of the winning Welsh pony mare, Danesfield Tina. Or Gaulden Gadabout, 12 h.h., by Solway Master Bronze (W.S.B.) out of Marilyn by Solway Golden Curlew out of Sulver Leaf. Master Bronze is by Coed Coch Glyndwr out of the show pony, Biddy Bronze, both Welsh; or the 14 h.h. Longmynd Golden Tango, by the Arab, Aloysha. These are only a few of the seemingly endless number of examples which could be quoted.

From the angle of colour, the most reliable but never certain crosses are Palomino with Palomino, or Palomino with chestnut. The result is usually either Palomino or chestnut. If the offspring of similar crosses are then re-crossed there is an even greater chance of getting a Palomino. Other colours, including greys and bays, do sometimes produce Palominos.

Palomino youngstock is often very pale at birth – a good fault, this, as the coats darken with the years. Indeed they also change surprisingly with the seasons, and it is for this reason that the rules and timing for inspection and registration are stated earlier in this chapter. Further details, lists of stallions, studs and so forth are available from the secretary.

Chapter 17

# THE QUANTOCK PONY: BRED IN THE WEST COUNTRY

The Quantock pony, like the Lundy Island pony, is a product of the West Country. He has a much longer history than his brother on Lundy Island, as ponies have been bred on the Quantock Hills in Somerset for many hundreds of years, although they have no Stud Book in which their breeding is recorded. They are very popular as riding ponies and can do almost any job, including playing polo and playing it well. Larger than many hill ponies some of them reach 14·2 h.h.

Roughly a hundred years ago these ponies were of almost pure Exmoor blood. Then it was decided to run good stallions carrying some Arab blood (although their dams have always been of mountain and moorland breeding), with the mares on the hills, which are collectively owned by farmers in this district. One of these was a grey stallion, Jude, an Arab-Welsh pony, full of quality. Another, the dark brown Waterfall, was by an Arab out of a New Forest mare. Before taking on his new duties he had won his class at one of the Ponies of Britain Stallion Shows.

The mares and foals are rounded up in September and driven to Nether Stowey, where they are penned and in due course the foals are separated, any fillies suitable for breeding being turned back on to the hills with their dams. This is done about two weeks before Bridgwater Fair, which is always held on the lst Wednesday in September, and the foals are on view at Nether Stowey for a few days before it takes place. Some are sold privately, others at the Fair. Quite a number have been exported to Scandinavia.

## Chapter 18

# THE SHETLAND PONY: VERY MUCH AN INDIVIDUAL

That so small and enchanting a woolly-bear pony as the Shetland should have played so useful a part in the life and development of his homeland, the Shetland Isles, is quite astonishing and certainly a credit to his durability and remarkable intelligence. There were no roads and little wheeled transport in Shetland until the middle of the last century, so goods, produce and people were carried by pony or else by boat.

These sturdy little individuals were essentially pack or riding ponies, whose origin is obscure but who roamed the hills of the Shetland Isles for hundreds of years – I believe ponies have existed there for 2,000 years – living in the open, regardless of weather conditions all the year round. Among their jobs was the carrying of seaweed for fertilizing the ground and this they would nibble and enjoy, no doubt benefiting by doing so.

For about a 100 of the 2,000 years mentioned above there was a demand in the coal pits for Shetland ponies and they were bred carelessly, even as far afield as Dartmoor, to meet this demand, many of the pit ponies being not of the best and true Shetland type, i.e. they were thick-necked, straight-shouldered and tubby. But the modern Shetland and those that were produced during that period by breeders of integrity and wisdom are indeed the true, strong pack and riding pony, and are ideal as a child's first mount. They are graceful as well as sure-footed, friendly, solid enough to give confidence and small enough to convince their small riders that the even more solid ground is not too far away!

Many people look upon the Shetland as the most intelligent of all the equine family, as they are very quick to learn – both good and bad things! They certainly love taking part in very varied activities and become extremely bored if not occupied either in breeding and rearing foals or being ridden, or played with. It is not enough for them to be just turned out in a field.

There have been and still are a number of great stallions, but one in particular, who will go down to history, is Jock, registered in the Stud Book under the number 16, and born in 1871. At this time the stock was becoming poor in quality and the Marquis of Londonderry established a stud on the Islands of Bressay and Ross. He owned Jock (16) and Jock had an immense and far-reaching influence on the breed, for which we are deeply indebted.

The Shetland pony's popularity has caused him to be widely exported. He is famous for the smallness of his stature combined with great strength and durability, like a miniature draught horse. But the large number of pedigree Shetlands exported from the Islands has depleted the stock from time to time.

As an example, during the years 1911–14, the Shetland Pony Stud Book Society issued about 500 export certificates to various countries abroad, principally the United States and Canada. There were, however, left on the Islands quite a few ponies who were pure-bred but not registered, and under these circumstances the Council decided to open the Stud Book to *inspected females only* belonging to members of the

Society, and about 700 were accepted for entry from 1911 to 1915. The produce from these females, sired by a registered, pedigreed stallion, were eligible for entry in the Stud Book.

For various reasons the war periods held up the breeding of pure Shetlands. Owing to the small demand for export and the loss of grazing through extra ploughing in the interests of food production, many breeders had to cut down their studs and omitted to keep proper records. The Society opened the Stud Book again for *inspected females* under similar conditions, and a large number have since been inspected and passed for entry. These have recently been entered in Section B of the Stud Book.

Under a scheme sponsored by the Shetland Pony Stud Book Society and the Department of Agriculture for Scotland, a number of carefully selected, pedigreed stallions were sent in 1948 to various locations in the Shetland Isles and this has been continued in subsequent years. This fresh infusion of blood was intended to improve the standard of breeding in the Islands and to encourage breeders to build up pedigreed studs again. I understand from Miss Nan French, who has kindly provided me with much information of the history of this breed, that this long-sighted, wisely thought-out scheme has proved, and is still proving, most successful.

In Holland are to be found as many, if not more, Shetland ponies than we have in the United Kingdom and only outstanding and very well-bred mares or stallions are imported there. France, Belgium, Sweden and Denmark are importing them, and indeed there is scarcely a country in the world to which Shetlands have not been sent at some time or other.

Here is the official description of the true Shetland:

HEIGHT; registered stock must not exceed 40 ins at three years old, and 42 ins at four years old and over. Ponies are measured from the top of the withers to the ground with a measuring stick on a level stance, preferably on concrete.

COLOUR; they may be black or any other colour known to horses.

COAT; the coat changes according to the seasons of the year – double coat in winter and smooth in summer.

HEAD; the head should be small and well-shaped, broad between the eyes and with small, well-placed ears. The eyes should be dark and intelligent.

NECK; this should have a good crest, especially in stallions; be strong and muscular and rise from a well-laid, oblique shoulder.

BODY; the body should have deep, well-sprung ribs, short back, with broad chest and strong quarters. The tail, like the mane, should be profuse and set fairly high. The loins should be strong and muscular.

LEGS; the forelegs should be well-placed under the shoulders, with well-muscled forearms and strong knees, followed by broad, flat bone and nice, springy pasterns. The hindlegs should have strong, muscular thighs, with broad, clean hocks, also broad, flat bone and springy pasterns as in the forelegs.

FEET; these should be open, round and tough.

ACTION; this must be free and true, with each joint freely used.

TRAINING; this little pony is responsive to firm, friendly treatment which he should receive from his earliest days. Always approach him, whether in the open or stabled, by first speaking to him (a rule which in fact applies to all horses and ponies) as a sudden, silent approach will startle him. This may lead to nervousness, mistrust and kicking – it is an instinctive reaction retained from the past when such an approach meant danger. Further training of the Shetland pony should follow that of the larger breeds and must be carried out equally carefully.

The secretary of the Shetland Pony Stud Book Society is Mr D. H. Patterson and his address is 8 Whinfield Road, Montrose, Angus. All stallions registered in the Society's Stud Book must now carry a Ministry of Agriculture licence.

The ponies fetch a high price at the official October sales, as much as 500 guineas having been paid recently for a filly at Aberdeen and 570 guineas at Lerwick. Very good classes are to be seen at some of the shows in Scotland and also at the Ponies of Britain summer show at Peterborough, Huntingdonshire, which features them to a considerable extent, as well as at the National Pony Show (this is not always held at the same venue), and at the Royal, the Royal Cornwall, the Devon County, the Bath and West, the East of England and the South of England shows.

This pony has long been driven in harness, both in the show-ring and for pleasure, and very charming he looks either singly or in pairs, to a show wagon or a neat little trap. In the United States the Shetland is no longer as we know him but has been transformed into a high-stepping, driving pony for the show-ring, and this is relevant to the question of first and subsequent crosses.

A breeder in Britain decided to cross Welsh Mountain mares with Shetland stallions and the resulting produce sold well in the States some years ago. There they were crossed in turn with small hackney ponies. This cross produced the steppers for driving and this is the type which is now popular.

Obviously show-ring driving is the only job for which these particular animals are suitable.

When judging in Canada in 1969 (not Shetlands!) an exhibitor talked to me of the alterations and improvements that he considered had been made to this breed in the States and in Canada, but although many of our breeds are thriving, retaining their character and producing progeny which is the greatest possible credit to those who breed and produce them in Canada and U.S.A., I could not agree in this instance. The Shetland has also been used in Canada to cross with bigger ponies of various types, sometimes with success, sometimes not, possibly because the mating was a trifle haphazard and not to any particular pattern.

By and large the Shetland is not used for crossing to any great extent in Britain. He is excellently suited to his own job, but not necessarily to produce any other type of pony, although no doubt there exist the notable exceptions to this rule.

## Chapter 19

# THE SPOTTED PONY AND HIS COLOURFUL PAST

Although the Spotted Pony is not at present listed among British native breeds one must not overlook the fact that there used to be an old strain of Spotted Ponies in Wales, that they have always aroused considerable interest, and that these showy little animals, as well as their bigger brothers, the Spotted Horses, are now being bred to quite an extent in Britain and are, I suppose, themselves an example of effective crossing.

The British Spotted Horse and Pony Society is endeavouring to establish a pure breed in this country and has a stud register dating from 1946. The secretary, Miss Joan Eddie of Nashend, Bisley, Stroud, Gloucestershire, will always supply names of breeders and of stallion owners and provide forms to register ponies if they are required.

The markings are as follows;

LEOPARD; spots of any colour on a white or light-coloured background.

BLANKET; a white rump or back on which there are spots of any colour.

SNOWFLAKE; white spots on a foundation of any colour.

Piebalds, skewbalds and dapple greys are not eligible. Naturally good quality is much looked for.

Typical characteristics of the breed consist of white sclera round the eye similar to the white round a human eye; striped hooves, yellowish white and black or brown in vertical stripes. The bare skin is mottled, usually under the dock, on the udder and round the eyes and muzzle. Manes and tails are often very sparse.

Mares who for some reason of marking or conformation do not quite come up to the standard required for the Main Register may enter the Grade Mare Register. If the mare breeds two or more good foals she may at a later date be accepted in the Main Register through the merits of her offspring.

Stallions are not entered in the Main Register until they hold a Ministry of Agriculture licence and have been passed by the Register Committee. Colts can be entered in the Colt Register, and on obtaining their licence will be transferred to the Main Register on payment of an extra fee of 20*s* (£1.00). If a colt does not get a Ministry of Agriculture licence by the time he is two years of age his papers will be cancelled. If gelded he can be transferred to the Gelding Register.

These ponies – and horses – date back to 1,800 B.C. and over the centuries they were to be observed in cave pictures, in Egyptian and Greek paintings and on a tomb wall in Northern Italy. They were domesticated in Asia about 3,500 years ago and later they were depicted in Chinese art as 'heavenly horses'. Although they were known in Arabia the colour is no longer found in Arab horses. The old Danish breed of Knabstruppers is exclusively confined to 'spotteds' and there are still stud farms in Denmark which feature the breed. They are also available in Holland and Sweden, where there is a spotted strain of Goathland pony.

## Chapter 20

# PONIES AND COBS OF WELSH STOCK

The designation 'Welsh Pony' covers a large, diverse and very admirable family, each with its own particular qualification, so let us begin by dividing it into its varous sections, as this will enable us to understand each type, their various attributes and what is expected of them.

Section A in the Welsh Stud Book is devoted to the Welsh Mountain pony, who must not exceed 12 h.h.

Section B is for the riding pony, not over 13·2 h.h.

Section C covers the pony of cob type, not over 13·2 h.h.

Section D is for cobs over this height.

Section E is for geldings (the progeny of an F.S.1 mare and a fully registered stallion – see below).

The description Foundation Stock or 'F.S.' can be explained as follows; F.S., approved ponies. F.S.1, one parent registered (if filly, F.S.2.; the colt, if gelded becomes Section E which is for geldings only). An F.S.1. mare may be put to a fully registered stallion and her filly is then F.S.2, her colt will be gelded and becomes Section E. If an F.S.2. is put to a fully registered stallion and produces a filly, the filly can be fully registered as an approved pony, i.e. Foundation Stock.

It is as well to have this information at one's finger tips

when purchasing stock, or considering the possibility of breeding or showing.

We will now look at each section in turn,

*The Welsh Mountain Pony* has a remarkable history in which its consistency of action, beauty and strength has helped it to accomplish not only remarkable feats but also to evoke pride and admiration. It is also claimed that the blood of this pony contributed in the very early days towards the breeding of the English Thoroughbred. The exact truth of this statement I do not know, but it is certain that it helped to produce the hackney, the polo pony, the hunter and unquestionably the Welsh cob.

It is obviously of the utmost importance that every care should be taken in the breeding of this beautiful little pony, and that reasons of expediency must never be allowed to override common sense and integrity – only the best must be accepted. The fact that there is such a demand for it may tempt the ill-advised breeder to hope he'll get by with the less-good pony, but in the long run this will be greatly to the detriment of the breed – and to his pocket and reputation.

The Welsh Mountain pony must have a swift, free action straight from the shoulder and he must use his hocks to the full. They should be well under the body, thus giving great propulsion and must be large, flat and clean, not turning out or in. The hind leg should not be too bent and the hock should not be set behind a line from the point of the quarter to the fetlock joint. Like any good riding pony, his pasterns are expected to be of proportionate slope and length and his feet well shaped.

The girth must be deep, ribs well sprung, back and loins strong and well coupled, hindquarters fine and of sufficient length to escape all criticism, i.e. not cobby, goose-rumped or ragged. His tail must be well set on and carried like a flag. The shoulders should be long and sloping well back, with moderately fine withers. The forelegs must not be set under the body nor tied in at the elbows; the forearm long, the knees well developed, with short, flat bone below.

The Welsh Mountain pony is full of character and courage, as one can observe from his bold eye, small clean-cut head, and sharp little ears, well up on the head. His nostrils are prominent and open, jaws and throat cleanly cut with plenty of room at the angle of the jaw, and his neck is well carried, sufficiently lengthy, and attractive. Naturally it is slightly cresty in the case of the stallion. Any colour is allowed, except piebald or skewbald.

Here is a fascinating pony who shows himself to advantage in the ring, whether ridden by a child, driven singly or in pairs – he certainly does not lack dash and his flowing mane and tail add glory to his way of going – or exhibited in in-hand classes.

It is therefore not surprising that the Welsh Mountain pony has been exported all over the world and that there are studs to be found as far from their native land as Queensland in Australia, where they are carefully bred and maintain their type, substance and beauty.

This breed is fortunate in having so many brilliant little stallions, and kindly, graceful, sturdy mares who throughout the past century have contributed so generously to the vital characteristics of their descendants, largely thanks to the wisdom and experience of those who owned the large herds or studs with their famous prefixes.

### The Welsh Pony (Section B)

This pony is summed up as follows by the Welsh Pony and Cob Society; 'The general description of ponies in Section A (Welsh Mountain, see previous pages) of the Stud Book is applicable to those *in* Section B, but more particularly the Section B pony shall be described as a riding pony, with quality, riding action, adequate bone and substance, hardness and constitution and pony character.' As we already know, he must not exceed 13·2 h.h.

He is a very lovely pony, ideal in children's show and performance classes particularly if his young rider's courage matches his own, as he has the heart of a lion. He is widely ridden in this type of class in Canada, and also out drag-hunting – he is natural jumper.

In Britain the best Section B young stock will often fetch very high prices. This stock is sometimes used for crossing with other breeds and with small Thoroughbreds or Arabs, according to what kind of pony is wanted. How a certain sire of unknown parentage made a great contribution to the breeding of Section B ponies is an unusual story.

This stallion, a Barb, was called Sahara, and carried what is known as the Sahara Desert brand, a necklace of spots in a double row round both sides of his shoulders and girth. His name often appears in pedigrees of Welsh ponies, largely because of his famous son, Tan-y-Bwlch Berwyn, his grandson, Coed Coch Berwynfa, and his various lovely grand-daughters.

Tan-y-Bwlch Berwyn was accepted into the Welsh Stud Book in 1927 at a time when new blood was needed. He was foaled in 1924 and grew to about 13 h.h. His sire is given as 'Sahara (847 National Pony Society's Stud Book), white Barb, 14·3 h.h.', and his dam was 'Brynhir Black Star, grey, 12 h.h., by the champion Shooting Star'.

Sahara had been purchased by Mrs F. E. G. Betty when she was in Gibraltar about 1913, out of a load of ponies from Morocco. Later, when World War I broke out he was shipped home with other horses as a charger, but as he was an entire he was not acceptable in this role and was sent to Mrs Betty in Wales, where he proved to be a very good hunter.

In the *Welsh Pony and Cob Society Journal* Mrs Betty tells us that eventually the local farmers induced her to use him for stud work and that he travelled the Carmarthenshire and Brecon districts for two seasons. She showed him as a polo pony stallion at the Royal, when it was held at Cardiff in 1919, and he did very well.

This is where Mr Denis Aldridge, for many years Secretary of the Quorn Hunt, spotted him and bought him, together with mares in foal to him. Both he and Mrs Betty speak with real affection of Sahara's kindness and quietness, Mr Aldridge declaring him to be the perfect hack.

The late Mrs Inge and the late Lord Lonsdale were among those who realized his value as a sire and sent a number of mares to him each year. All those sent by Mrs Inge were grey Welsh ponies and among them was the mare who became the dam of Tan-y-Bwlch Berwyn. Mr Aldridge says: 'No mating has ever done more for the pony breeds than this. Sahara was the surest of stock getters and I never remember his getting anything but a grey.' Incidentally he also sired some very valuable polo ponies.

There are today a number of studs which specialize in the pure Welsh Mountain pony and also in the Section B ponies, crossing the latter in their efforts to produce what they believe to be the ideal riding pony.

### The Welsh Pony of Cob Type and The Welsh Cob ((Sections C and D)

Of all horses and ponies one of man's greatest friends is the Welsh cob, practical, intelligent and with just that presence and dash which add zest to everyday life. He is a most self-respecting animal who will do his best for his owner strictly on a partnership basis not as a servant, provided no-one ever tries to make a fool of him.

Today the tragedy is that so much of the good cob stock has been allowed to waste and although there are still a few good animals available for breeding the right type there is not only a certain lack of agreement as regards what constitutes the right type, and the right action, but so many breeders seem to think that ponies are a more profitable speculation than cobs. In spite of this, they are becoming more and more appreciated and an increasingly successful sale of cobs is held annually, thanks to the foresight and energy of Miss Pauline Taylor at her stud, the Llanarth Stud, Blaenwern, Llanarth, Cardiganshire. This is an open sale, not confined to her own cobs and ponies, and as a result many good cobs and ponies are finding homes both in England and abroad.

The best type of Welsh cob is an excellent ride, handy as a polo pony and a very fine hunter, particularly over trappy country. He has a wise eye for a difficult obstacle and a natural gift for jumping, always ready to jump off his hocks. He can gallop on with the best but does not aspire to compete with the Thoroughbred when it comes to speed – for all that he'll make sure you are there at the finish of a good run.

A cob has the pleasant gift of companionship, will be able to work out things for himself and is prone to have decided preferences. One of the best horses I ever had was a black Welsh cob, and his behaviour during the air raids on London in World War II evoked the highest respect and admiration. In addition to this there was little he could not do if he chose and he had great courage and kindness. As his valediction I would say that his cobby character and good sense left some human beings way behind. As far as I remember he came originally from a sale at or near Bridgnorth and I had him when he was four years old. He died at about 26 after a great life of service.

Here is the official description of the Welsh Pony of cob type, which must not exceed 13·2 h.h., and also of the Welsh cob.

GENERAL CHARACTER; sturdy, strong and active, with pony character and as much substance as possible.

COLOUR; any colour except piebald and skewbald.

HEAD; full of quality and pony character. A coarse head and Roman nose are most objectionable.

EYES; bold, prominent and set widely apart.

NECK; lengthy and well-carried. Moderately lean in the case of mares, but inclined to be cresty in the case of mature stallions.

SHOULDERS; strong but well laid back.

FORELEGS; set square and not tied in at the elbows. Long, strong forearms. Knees well-developed with an abundance of bone below them. Pasterns of proportionate slope and length. Feet well shaped. Hoofs dense. When in the rough, a moderate amount of silky feather is not objected to but coarse, wiry hair is a definite objection.

MIDDLEPIECE; back and loins muscular, strong and well-coupled. Deep through the heart and well ribbed-up.

HINDQUARTERS; lengthy and strong. Ragged or drooping quarters are objectionable. Tail well set on.

HIND LEGS; second thighs strong and muscular. Hocks large, flat and clean, with points prominent, turning neither inwards nor outwards. The hind legs must not be too bent and the hock not set behind a line falling from the point of the quarter to the fetlock joint. Pasterns of proportionate slope and length. Feet well shaped. Hoofs dense.

ACTION; free, true and forcible. The knee should be bent and the whole foreleg should be extended straight from the shoulder and as far forward as possible in the trot. Hocks flexed under the body with straight and powerful leverage.

One does not see as large a number of *good* cobs competing in their own class as one would wish, and in the mixed events for mountain and moorland breeds it is sometimes difficult to accede to them the honour which is their due. Quite often Sections C and D (small and large) are judged together and this usually bears out the truth of the old saying that 'a good big "un" will always beat a good little "un"', though sometimes the decision is questionable.

At one time these remarkably useful, showy and active ride and drive animals were bred with great care and pride in Wales, but today, alas, their numbers and quality as a whole have diminished. Thanks to the work of one or two excellent studs this situation is likely to improve.

# USEFUL ADDRESSES

Whether you decide to breed an odd pony or two, start a stud or even extend one you already have, you may well be interested in the examinations in stud work held by the National Pony Society. These cover simple, practical and also more advanced stud work and are designed to help the individual who wishes to work in a stud without actually riding, as well as the one who will include riding in his or her curriculum. Both diplomas and stud certificates are awarded, and they are useful to anyone who wishes to learn stud work and make their career this way and also to the prospective employer of stable staff. A study of the synopses is most rewarding and well worth the trouble of writing to the secretary of this society, Commander Brown, R.N., 85 Cliddesden Road, Basingstoke, Hants. Will readers in the U.K. please enclose a postage stamp for reply.

A very practical suggestion, if you lack experience, is to work at a good, recognized stud before starting one of your own, as this will give you not only the experience you need but an idea of costs. To breed ponies successfully it is necessary to buy or rent suitable land, though fortunately its proximity to London is not essential, except for your own amusement.

But what is of the utmost importance is to know how to keep – and balance – accounts meticulously, as well as deal promptly and accurately with correspondence, pedigrees, registration and so forth. And you must be one of those lucky people who have an eye – and an ear – for the current and future market plus the knack of breeding what is wanted.

Your ponies can be bought and sold through recommendations, by advertising, or at the various pony sales such as Fayre Oakes, the National Pony Sales and others, including breed sales, which take place regularly. The larger ones are all advertised. In the case of the others the breed societies concerned will usually inform you of the probable annual dates if you write to them.

Here are the addresses of the some of the secretaries and of the various societies. You already have that of the National Pony Society, which publishes each year a most interesting illustrated review, containing a large number of excellent articles and advertisements.

CONNEMARAS

English Connemara Pony Society, Mrs Barthorp,
The Quinta, Bentley, Farnham, Surrey.

DALES

Dales Pony Society, George H. Hodgson,
Ivy House Farm, Hilton, Yarm-on-Tees, Yorkshire.

DARTMOORS

Dartmoor Pony Society, Dr E. L. Roberts,
Lower Hisley, Lustleigh, Newton Abbot, S. Devon.

EXMOORS

Exmoor Pony Society, Lt-Col V. C. A. Munckton,
Capland Orchard, Hatch Beauchamp, Taunton, Somerset.

FELLS

Fell Pony Society, Miss P. Crossland,
Packway, Windermere, Westmorland.

HIGHLANDS
Highland Pony Society,
51 High Street, Dunblane, Perthshire.

NEW FOREST PONIES
New Forst Pony Breeding and Cattle Society,
Miss D. Macnair,
Beacon Corner, Burley, Ringwood, Hants.

SHETLANDS
Shetland Pony Stud-Book Society, Mr D. M. Patterson,
8 Whinfield Road, Montrose, Angus, Scotland.

WELSH PONIES AND COBS
Welsh Pony and Cob Society,
32 North Parade, Aberystwyth, Cardiganshire.

PALOMINOS
British Palomino Society, Mrs Peter Howell,
Kingsettle Stud, Cholderton, Salisbury, Wiltshire.

SPOTTED PONIES
British Spotted Horse and Pony Society, Miss Joan Eddie,
Nashend, Bisley, Stroud, Gloucestershire.

PONIES OF BRITAIN
Ponies of Britain Club, Mrs Glenda Spooner,
Brookside Farm, Ascot, Berkshire.

# INDEX

# INDEX

# INDEX